D1709051

IMPACTS OF CLIMATE CHANGE ON ALLERGENS AND ALLERGIC DISEASES

Climate change has been identified as the biggest global health threat of the twenty-first century (*The Lancet*). Hundreds of millions of people around the world currently suffer from allergic diseases such as asthma and allergic rhinitis (hay fever), and the prevalence of these diseases is increasing. This book is the first authoritative and comprehensive assessment of the many impacts of climate change on allergens, such as pollen and mould spores, and allergic diseases. The international authorship team of leaders in this field explore the topic to a breadth and depth far beyond any previous work. This book will be of value to anyone with an interest in climate change, environmental allergens, and related allergic diseases. It is written at a level that is accessible for those working in related physical, biological, and health and medical sciences, including researchers, academics, clinicians, and advanced students.

PAUL J. BEGGS is Professor in the Department of Environmental Sciences at Macquarie University, Australia. In 2009 he was awarded the Eureka Prize for Medical Research for his research on the impacts of climate change on allergens and allergic diseases. He was a contributing author of the *Fourth Assessment Report of the Intergovernmental Panel on Climate Change*, published in 2007, the same year the Intergovernmental Panel on Climate Change won the Nobel Peace Prize. He was the President of the International Society of Biometeorology from 2008 to 2011.

IMPACTS OF CLIMATE CHANGE ON ALLERGENS AND ALLERGIC DISEASES

Edited by

PAUL J. BEGGS
Macquarie University

CAMBRIDGE
UNIVERSITY PRESS

University Printing House, Cambridge CB2 8BS, United Kingdom

Cambridge University Press is part of the University of Cambridge.

It furthers the University's mission by disseminating knowledge in the pursuit of education, learning and research at the highest international levels of excellence.

www.cambridge.org
Information on this title: www.cambridge.org/9781107048935

First published 2016

Printed in the United States of America by Sheridan Books, Inc.

A catalogue record for this publication is available from the British Library.

Library of Congress Cataloguing in Publication Data
Names: Beggs, Paul J., 1968– editor.
Title: Impacts of climate change on allergens and allergic diseases / Paul J. Beggs, editor, Department of Environmental Sciences, Faculty of Science and Engineering, Macquarie University.
Description: Cambridge : Cambridge University Press, 2016. | Includes bibliographical references and index.
Identifiers: LCCN 2016017601 | ISBN 9781107048935 (hardback)
Subjects: LCSH: Climatic changes – Health aspects. | Environmental health. | Allergens. | Respiratory allergy.
Classification: LCC QC903.I448 2016 | DDC 614.5/993–dc23
LC record available at https://lccn.loc.gov/2016017601

ISBN 978-1-107-04893-5 Hardback

This book is dedicated to Dr. Diana J. Bass, who in the early 1990s generously introduced me to the wonderful world of airborne pollen and mould spore monitoring and, in so doing, set me on the path that would ultimately lead to the production of this book.

There are likely to be quantitative and/or qualitative changes in the airborne concentration of allergens, e.g., molds and pollens. This in turn could lead to changes in the prevalence or intensity of asthma and hay fever episodes in affected individuals.

(Janice Longstreth, "Anticipated public health consequences of global climate change")

Contents

Colour plates are to be found between pages 136 and 137.

Figures

Tables

Contributors

Linda J. Beaumont, PhD
Department of Biological Sciences
Faculty of Science and Engineering
Macquarie University
New South Wales 2109
Australia

Paul J. Beggs, PhD
Department of Environmental Sciences
Faculty of Science and Engineering
Macquarie University
New South Wales 2109
Australia

Jeroen T. M. Buters, PhD
Center of Allergy and Environment (ZAUM)
Technical University Munich and Helmholtz Center Munich
Biedersteiner Strasse 29
80802 Munich
Germany
and
Christine Kühne Center for Allergy Research and Education
Davos
Switzerland

Ginger L. Chew, ScD
Centers for Disease Control and Prevention
National Center for Environmental Health
Division of Environmental Hazards and Health Effects

Air Pollution and Respiratory Health Branch
4770 Buford Hwy., N.E., MS-F60
Atlanta, GA 30341
United States of America

Daisy E. Duursma, MSc
Department of Biological Sciences
Faculty of Science and Engineering
Macquarie University
New South Wales 2109
Australia

Susanne Jochner, PhD
Physical Geography / Landscape Ecology and
Sustainable Ecosystem Development
Catholic University of Eichstätt-Ingolstadt
Ostenstraße 18
85072 Eichstätt
Germany

Constance H. Katelaris, PhD, MBBS
School of Medicine
Western Sydney University
Campbelltown 2560
New South Wales
Australia
and
Immunology and Allergy
Department of Medicine
Campbelltown Hospital
Campbelltown 2560
New South Wales
Australia

Patrick L. Kinney, ScD
Climate and Health Program
Mailman School of Public Health
Columbia University
722 West 168th Street
New York, NY 10032
United States of America

Annette Menzel, PhD
Ecoclimatology
Technische Universität München
Hans-Carl-von-Carlowitz-Platz 2
85354 Freising
Germany

Rachel L. Miller, MD
Division of Pulmonary, Allergy, and Critical Care Medicine
Columbia University Medical Center
PH8E-101B, 630 W.
168th Street
New York, NY 10032
United States of America

Marje Prank, MSc
Finnish Meteorological Institute
Erik Palménin aukio 1
FI-00560 Helsinki
Finland

Shubhayu Saha, PhD
Centers for Disease Control and Prevention
National Center for Environmental Health
Division of Environmental Hazards and Health Effects
Air Pollution and Respiratory Health Branch
4770 Buford Hwy., N.E., MS-F60
Atlanta, GA 30341
United States of America

Mikhail Sofiev, PhD
Finnish Meteorological Institute
Air Quality Research
Erik Palménin aukio 1
FI-00560 Helsinki
Finland

Kate R. Weinberger, PhD
Institute at Brown for Environment & Society
Brown University
Box 1951

85 Waterman Street
Providence, RI 02912
United States of America

Lewis H. Ziska, PhD
Crop Systems and Global Change
Agricultural Research Service
US Department of Agriculture
10300 Baltimore Avenue
Beltsville, MD 20705
United States of America

Preface

This book considers the impacts of climate change on allergens and allergic diseases. It is the first book to focus on this topic. It provides a comprehensive and up-to-date review, assessment, and synthesis of this topic based on the scientific literature. In addition, two of the chapters (Chapters 4 and 7) also present new findings.

Warming of the climate system is unequivocal, and the concentrations of greenhouse gases such as carbon dioxide have increased. These and many other conclusions by the Intergovernmental Panel on Climate Change's Working Group I in its contribution to the Fifth Assessment Report published in 2013 provide the climate change context for this book. The introductory chapter (Chapter 1) provides a brief description of this, focusing on aspects of climate change most relevant to allergens and allergic diseases.

The book considers both observed (past and current) and projected (future) impacts. The spatial scope of the book is global and international. However, the nature of this topic requires that the full range of scales be considered, from the micro and molecular to the macro.

The book consists of ten chapters. Seven of these (Chapters 2 to 8) focus primarily on the impacts of climate change on allergens per se. Each of these chapters explores a different aspect: aeroallergen production and atmospheric concentration (Chapter 2); the distributions of allergenic species (Chapter 3); aeroallergen dispersion, transport, and deposition (Chapter 4); allergenicity (Chapter 5); allergen seasonality (Chapter 6); indoor allergens (Chapter 7); and interactions among air pollutants and aeroallergens (Chapter 8). Chapter 9 explores climate change impacts on allergic diseases explicitly. A synthesis of the preceding nine chapters and an overview of mitigation and adaptation responses in the context of climate change impacts on allergens and allergic diseases are presented in Chapter 10. This final chapter also highlights a range of knowledge gaps and research needs. An impressive list of allergen-producing organisms, allergens, and allergic diseases is discussed. The former include a wide range of plants (trees, shrubs and weeds, and grasses), fungi, cockroaches, house dust mites, mice, and stinging insects. The allergic diseases considered here range from asthma and allergic rhinitis to allergic rhinoconjunctivitis, atopic dermatitis, insect sting allergy, and food allergy. Much of the focus, however, is on plants and the pollen they produce, and asthma and allergic rhinitis.

The chapters of this book have been contributed by fifteen authors in all. The lead authors of Chapters 2 to 9 are internationally acclaimed experts and have been specially invited to write on their respective subjects. Most of the lead authors have preferred to take on one or more of their colleagues as co-authors. The authorship team has represented both the Northern and Southern Hemispheres, three geographical regions (Europe, Northern America, and Oceania), and four countries (Australia, Finland, Germany, and the USA). They also represent a range of institutions, including universities, a national disease control and prevention centre, hospitals and medical centres, a national meteorological institute, and a national agricultural research service. Each of the authors has approached the given subject from their own disciplinary perspective: ecology, environmental health sciences, allergy and immunology, meteorology, botany, and so on. However, the experience and expertise of each of the authors transcends any one discipline; they are all truly and necessarily interdisciplinary. The introductory and concluding chapters (Chapters 1 and 10) have been written by the book's editor, with Chapter 10 being co-authored by the lead author of Chapter 6.

The text of the book is complemented with tables and figures where appropriate, and cross-referencing between chapters enhances integration and minimises duplication. However, different aspects of a topic cannot be properly considered in total isolation and so some overlap is necessary and desirable. Finally, a broad range of acronyms and abbreviations are used in the book, and a consolidated list is provided in this front matter.

Ultimately, this book presents an authoritative picture of what we know about the impacts of climate change on allergens and allergic diseases, and a call for action – appropriate responses to what we know, and more research to fill the gaps in our knowledge.

Acknowledgements

I would first like to thank all the authors of this book. Their acceptance of my invitation to contribute to this book in many respects surprised and overwhelmed me. Some I had never met and only knew by their reputation, yet they generously embarked on this long journey with me. While all chapters presented a challenge to their author(s), some were more challenging because I had requested them to write on an aspect of this topic that, while important, had received little if any attention previously. And in two particular cases, the authors' progress on the writing suffered a setback when they encountered monumental personal hardships, but both remained loyal and committed to the project and came through with their chapters.

A special thank you also to Dr Matt Lloyd, Publishing Director, Science, Technology, and Medicine, Americas, at Cambridge University Press. Again, this book would not have existed without Matt. His visit to Macquarie University at the end of July 2012, for which he had prearranged a meeting with me to discuss any book idea I might have, enabled me to pitch the idea to him, and his immediate enthusiasm and encouragement resulted in me getting a book proposal in to Cambridge University Press just a month later. Matt and I have only met once, but his guidance and support through this process has been unwavering, even when the project continued on for much longer than I had thought (although I am sure he has seen it all before, and hopefully he has had book projects that took longer than mine!). Beyond Matt taking the initiative to meet with me back in 2012 and his provision of sage advice at a number of crucial times, I thank him for providing me with the freedom and flexibility to bring things together as and when I could. I took on the position as Head, Department of Environmental Sciences at Macquarie, in January 2013, and as much as I made this book project a priority, the challenges and demands of this position likely meant some things took longer than they otherwise would have. I really appreciate Matt for not pressuring me during that time.

I also thank the many others at Cambridge University Press who have contributed to the production of this book in one way or another. In particular, I thank

David Morris (Project Manager, Academic Publishing), Cassi Roberts (Content Assistant, Academic Books), and Zoë Pruce (Content Manager, Academic Books) for expertly guiding me through the production process. I also acknowledge the amazing work of Ramesh Karunakaran (Project Manager) and Pradeep Kumar (Copy Editor) and others at Newgen during the copyediting, typesetting, and proofing stages.

As part of the process of putting this book together, I had sent chapters to reviewers for comment and feedback. They too have made a valuable contribution to the quality of this book, and I thank them for this: Lorenzo Cecchi (Università degli Studi di Firenze (University of Florence, Italy); Bernard Clot (Federal Office of Meteorology and Climatology MeteoSwiss, Switzerland); Kris Ebi (University of Washington, USA); Simon Haberle (The Australian National University, Australia); and Mark Schwartz (University of Wisconsin-Milwaukee, USA). And I thank two anonymous reviewers of the original book proposal submitted to Cambridge University Press.

And finally I wish to acknowledge all those who have mentored, encouraged, and supported me over the many years of my career, and prior to that. Thank you to Macquarie University and my many colleagues and friends there. Similarly, thank you to my colleagues and friends in the wonderful and intriguing worlds of biometeorology, aerobiology, and environmental health science. What better professional communities could an academic and scholar hope for? And to my partner Leanne (and our dog Caley), my family, and my friends – thanks is simply not enough, but thank you anyway.

Acronyms and Abbreviations

AAAAI	American Academy of Allergy, Asthma, and Immunology
ABL	Atmospheric boundary layer
AC	Air-conditioning
AC	Allergic conjunctivitis
AD	Atopic dermatitis
AHS	American Housing Survey
AP	Allergen-producing
API	Annual pollen index
AR	Allergic rhinitis
A_w	Water activity
β2AR	β2 adrenergic receptor
CCCEH	Columbia Center for Children's Environmental Health
CDC	Centers for Disease Control and Prevention (USA)
CEE	Central and Eastern Europe
cfu	Colony-forming unit
CH_4	Methane
CI	Confidence interval
CO	Carbon monoxide
CO_2	Carbon dioxide
DALYs	Disability-adjusted life years
DC	District of Columbia
DEPs	Diesel exhaust particles
DNA	Deoxyribonucleic acid
DOY	Day of the year
EAN	European Aeroallergen Network
EC	Elemental carbon
ECMWF	European Centre for Medium-Range Weather Forecasts

ECRHS	European Community Respiratory Health Survey
ED	Emergency department
EPA	Environmental Protection Agency (USA)
FACE	Free-air CO_2 enrichment
FEMA	Federal Emergency Measurement Agency (USA)
FIA	Forest Inventory and Analysis
GA^2LEN	Global Allergy and Asthma Network of Excellence
GCMs	General circulation models
GHG	Greenhouse gases
GINA	Global Initiative for Asthma
GIS	Geographic information systems
GM	Geometric mean
HDM	House dust mite
HIALINE	Health Impacts of Airborne Allergen Information Network
IFN-γ	Interferon gamma
IgE	Immunoglobulin E
IL-4	Interleukin 4
IL-10	Interleukin 10
IPCC	Intergovernmental Panel on Climate Change
ISAAC	International Study of Asthma and Allergies in Childhood
KLH	Keyhole limpet hemocyanin
K_{z1}	Turbulent diffusion coefficient at 1 m above the ground
LPG	Liquefied petroleum gas
MAS	Multicenter Allergy Study
MSAs	Metropolitan statistical areas
NHANES	National Health and Nutrition Examination Survey (USA)
NO_2	Nitrogen dioxide
NOAA	National Oceanic and Atmospheric Administration (USA)
NO_x	Reactive nitrogen oxides (sum of NO and NO_2)
O_3	Ozone
OC	Organic carbon
OR	Odds ratio
OVA	Ovalbumin
PAHs	Polycyclic aromatic hydrocarbons
PAR2	Protease-activated receptor 2
PCR	Polymerase chain reaction
PEFR	Peak expiratory flow rate
PM	Particulate matter
$PM_{2.5}$	Particulate matter with aerodynamic diameter <2.5 μm

PM_{10}	Particulate matter with aerodynamic diameter <10 μm
ppb	Parts per billion
ppm	Parts per million
PSP	Polystyrene particles
Q_2	Relative humidity at 2 m above the ground
RCP	Representative concentration pathway
RH	Relative humidity
SARC	Seasonal allergic rhinoconjunctivitis
SDM	Species distribution model
sIgE	Specific immunoglobulin E
SILAM	System for Integrated modeLing of Atmospheric composition
SO_2	Sulphur dioxide
TENOR	The Epidemiology and Natural History of Asthma: Outcomes and Treatment Regimens
Th1	Type 1 T helper cell
Th2	Type 2 T helper cell
τ_{pr}	Fraction of rainy periods per month
u	Longitudinal wind component
U_{10}	Wind speed at 10 m above the ground
v	Latitudinal wind component
VOCs	Volatile organic compounds
WHO	World Health Organization
ϕ_{10}	Wind direction at 10 m above the ground

1

Introduction

PAUL J. BEGGS
Department of Environmental Sciences
Faculty of Science and Engineering
Macquarie University

1.1 Introduction

Climate change is *the* issue of our time. It is global, international, and pervasive. Of all the impacts of climate change, those on human health are perhaps the most significant. Indeed, the prestigious medical journal *The Lancet* recently stated that 'Climate change is the biggest global health threat of the 21st century' (Costello *et al.*, 2009).

The impacts of climate change on human health are many and varied. Beyond what are thought of as the direct impacts on human health, such as the direct effects of temperature extremes and severe weather, are a multitude of indirect impacts of climate change on human health, or what Butler (2014) has recently described as secondary (and tertiary) effects. The UN Intergovernmental Panel on Climate Change (IPCC) has most recently described these indirect or secondary impacts on human health as 'ecosystem-mediated impacts' (Smith *et al.*, 2014). The impacts of climate change on allergic diseases fall clearly within this realm.

Allergic diseases, such as asthma and allergic rhinitis, are of global importance for a number of reasons. It is estimated that 235 million people currently suffer from asthma, this being the most common non-communicable disease among children (World Health Organization, 2015). The prevalence of allergic diseases has increased dramatically over recent decades and continues to increase (Pearce *et al.*, 2007). And allergic disease markedly affects the quality of life of both individuals with this disease and their families and negatively impacts the socioeconomic welfare of society (Pawankar *et al.*, 2011).

Our environment contains allergens from many sources. These include pollen from trees, weeds and grasses, mould spores, house dust mites, cockroaches, and others. Climate plays a major role in the lives of allergenic organisms, as well as their production of allergens and our eventual exposure to such allergens. Climate influences the distribution and abundance of all allergenic organisms. Similarly, the variations in temperature, precipitation, humidity, and other factors that characterise the seasons control the activities of allergenic organisms, including their production of allergens. The so-called pollen season is perhaps the best known example of this. Climate extremes are also important, with the rampant growth of mould indoors following flooding of buildings, such as that in New Orleans following Hurricane Katrina, and the phenomenon of 'thunderstorm asthma' being just two examples of this. It is therefore to be expected that climate change would result in changes to allergenic organisms, exposure to their allergens, and allergic diseases.

The impacts of climate change on allergens and allergic diseases have progressively received increasing attention over the last 25 years or so, both as a topic and as an issue. In particular, the impacts of climate change on aeroallergens and allergic respiratory diseases were highlighted as one of only seven key health effects that supported the US Environmental Protection Agency's (EPA) finding that current and future concentrations of greenhouse gases endangered public health, in 'Endangerment and Cause or Contribute Findings for Greenhouse Gases under the Clean Air Act' (US EPA, 2009, 2016). Such impacts were the focus of one of just eight chapters on climate change health effects in the recent *Health Effects of Climate Change in the UK 2012* report (Kennedy and Smith, 2012). Perhaps most recently, the topic has again received prominent attention, being the focus of a chapter in the book titled *Climate Change and Global Health* (Beggs, 2014).

There has been much outstanding research on this topic – most prominent among this were the study by Ziska *et al.* (2011) published in the *Proceedings of the National Academy of Sciences of the United States of America* demonstrating lengthening of the ragweed pollen season in North America in recent decades due to warming over this period, and a very recent study by Hamaoui-Laguel *et al.* (2015) published in *Nature Climate Change* showing that airborne ragweed pollen concentrations in Europe will be approximately four times higher by 2050 than they currently are as a result of future climate and land use changes. The acceleration of research in this area has been astounding, with, for example, one recent analysis showing that since 1998, one-third of the literature on this topic has been published in just a span of two-and-a-half years, from 2013 to mid-2015 (Beggs, 2015).

The topic is now at a turning point. What is needed is a comprehensive and authoritative assessment of the whole of this topic to clearly document where we stand in terms of our understanding of this topic and to highlight gaps in our knowledge and research priorities for the future. This book, the first one to be entirely devoted to the impacts of climate change on allergens and allergic diseases, aims to fill this need. The following section provides a brief description of climate change itself – the changes in the composition of the Earth's atmosphere, its temperature, precipitation, and so on. The final section introduces the other nine chapters of this book.

1.2 Climate Change

Our climate is changing. Climate change due to human activities, the focus of this book, has been witnessed for at least the last 100 years and is projected to continue for centuries to come. Climate change involves the whole climate system, including not only our atmosphere but also our hydrosphere, cryosphere, land surface, and biosphere. With other authoritative and in-depth assessments of climate system changes readily available (i.e., IPCC, 2013a), the purpose of this section is to provide a brief description of 'the physical science basis' of climate change, focussing on the aspects of climate change most relevant to allergens and allergic diseases.

The atmospheric concentrations of several greenhouse gases have increased since the start of the Industrial Era (1750). Atmospheric carbon dioxide (CO_2) concentrations have increased by 41% since this time, primarily from fossil fuel emissions and secondarily from net land use changes (IPCC, 2013b). The most recent global annual mean atmospheric CO_2 concentration, for 2013, was 395.22 parts per million (ppm) (National Oceanic and Atmospheric Administration (NOAA), 2015a), an increase of over 100 ppm from the pre-Industrial Era value of approximately 280 ppm. As the records from the Mauna Loa Observatory illustrate (Figure 1.1), the increase in atmospheric CO_2 concentration since 1750 has not been linear, with much of the increase occurring in just the last 60 years or so and the increase during this last 60 years getting steeper and steeper toward the present time.

This increase in the atmospheric concentration of greenhouse gases such as CO_2 has led to an uptake of energy by the climate system (IPCC, 2013b), and this has resulted in observed warming of the climate system. Between 1880 and 2012, the Earth's average surface temperature warmed by 0.85°C (with a 90% confidence interval (CI) of 0.65°C–1.06°C) (IPCC, 2013b). Most of this warming (0.72°C (CI 0.49°C–0.89°C)) occurred after 1951 (Hartmann *et al.*, 2013). Warming of the Earth's surface has also varied over space, with, for example, the land surfaces tending to warm more than the oceans. This means that some parts of the Earth's surface have warmed considerably more than the average of 0.85°C, as much as double or more in some places.

Changes in precipitation have also been observed. For example, since 1901, precipitation has increased over the mid-latitude land areas of the Northern Hemisphere (Hartmann *et al.*, 2013). Other components of the Earth's hydrological cycle have also changed. The moisture content of the air around us, and in our

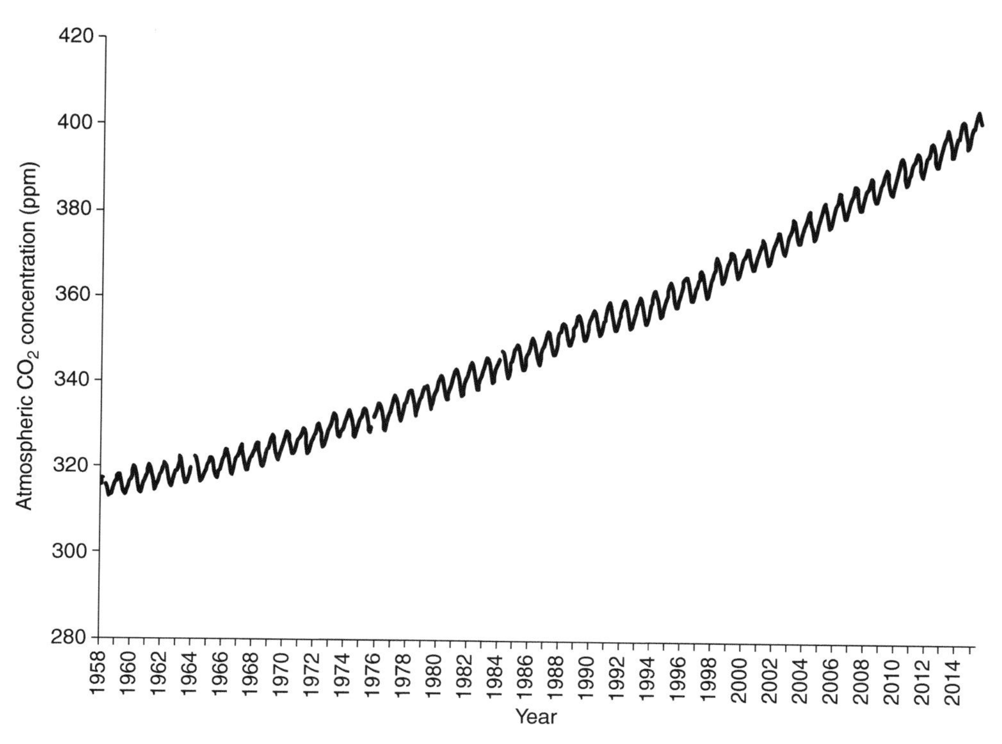

Figure 1.1. Monthly mean atmospheric carbon dioxide concentration at Mauna Loa Observatory from March 1958 to July 2015. Note that the last year of data is preliminary. A few months are missing in the records and these account for the gaps in the graph.
Source: NOAA (2015b).

environment generally, has increased since the 1970s. Finally, it is likely that the Earth's general atmospheric circulation has changed, and in particular such features as storm tracks and jet streams have moved poleward since the 1970s, involving a widening of the tropical belt and a contraction of the northern polar vortex (Hartmann *et al.*, 2013).

The past half century or so has also been assessed to have experienced changes in a range of extreme weather and climate events. In terms of temperature extremes, over most land areas, such changes include warmer and/or fewer cold days and nights, warmer and/or more frequent hot days and nights, and increased frequency and/or duration of warm spells/heat waves (IPCC, 2013b). In terms of precipitation extremes, on the one hand, there has been an increase in the frequency, intensity, and/or amount of heavy precipitation, and on the other hand there have been increases in intensity and/or duration of drought (IPCC, 2013b). And finally, intense tropical cyclone (hurricane) activity has increased. More fundamentally though, Trenberth (2012) concludes that 'all weather events are affected by climate change because the environment in which they occur is warmer and moister than it used to be'. He supports this conclusion with an eloquent commentary on the climate system, climate change, and recent climate extremes.

As significant and important these observed changes in the climate system are, they are only half the picture. To complete the picture, we must also look into the future. This requires information about future emissions or concentrations of greenhouse gases, aerosols, and other climate drivers, and the scientific community has developed sets of potential scenarios (of human activities and corresponding emissions, etc.), the latest being labelled Representative Concentration Pathways (RCPs).

Sophisticated climate models that have evolved over decades of development provide us with a range of possible climate futures based on the RCPs. Such possible future climates are referred to as climate change projections. According to the IPCC, 'A projection is a potential future evolution of a quantity or set of quantities, often computed with the aid of a model. Unlike predictions, projections are conditional on assumptions concerning, for example, future socioeconomic and technological developments that may or may not be realized' (Agard *et al.*, 2014; IPCC, 2013c).

There are four RCPs (labelled RCP2.6, RCP4.5, RCP6.0, and RCP8.5), according to which atmospheric CO_2 concentrations will reach 421, 538, 670, and 936 ppm by 2100, respectively (IPCC, 2013b). The first pathway (RCP2.6) is thought of as a 'mitigation scenario' (a topic to be expanded upon in Chapter 10 of this book). The second and third pathways are thought of as 'stabilisation scenarios', and the fourth pathway is one with very high greenhouse gas emissions.

By the end of this century (2081–2100), global mean surface temperatures are projected to increase relative to those for the period 1986–2005. The Earth's surface temperature is projected to increase anywhere from 0.3°C to 4.8°C, wherein the extent of warming depends on the pathway: 0.3°C to 1.7°C (RCP2.6), 1.1°C to 2.6°C (RCP4.5), 1.4°C to 3.1°C (RCP6.0), and 2.6°C to 4.8°C (RCP8.5) (IPCC, 2013b). Regardless of the pathway followed, this further warming will not be uniform across the surface of the Earth. The mean warming over land will be larger than over the ocean, and the Arctic region will warm more rapidly than the global mean (IPCC, 2013b).

Globally, on average, precipitation is projected to increase by the end of this century (Collins *et al.*, 2013). However, there will be substantial spatial variation in precipitation changes, with some regions experiencing increases, some decreases, and some no change at all. The IPCC has concluded: 'that the contrast of annual mean precipitation between dry and wet regions and that the contrast between wet and dry seasons will increase over most of the globe as temperatures increase' (Collins *et al.*, 2013). Atmospheric moisture will generally increase into the future.

Complex changes in atmospheric circulation are projected. These include an increase in the area encompassed by monsoon systems and a lengthening of the monsoon season in many regions. Monsoon winds, however, will likely weaken (IPCC, 2013b). Similarly, the Hadley and Walker Circulations in the tropics are likely to slow down (Collins *et al.*, 2013).

Climate extremes will also continue to change into the future. Hot temperature extremes, including heat waves, will be more frequent, and cold temperature extremes will be less frequent, over most land areas. Extreme precipitation events will become more intense and more frequent over most of the mid-latitude land masses and over wet tropical regions (IPCC, 2013b).

1.3 The Chapters that Follow

The chapters that follow discuss the impacts of climate change on allergens and allergic diseases under eight sub-topics, with the final chapter acting to provide a synthesis of these eight sub-topics and a conclusion to the book as a whole. As is made clear in this introductory chapter, each of the following chapters will consider both observed (past and current) and projected (future) impacts, where possible. Similarly, the spatial scope of the book is global and international. However, the nature of this topic requires that the full range of scales be considered, from the micro and molecular to the macro.

Chapter 2 focusses on the impacts of climate change on aeroallergen (pollen and fungal spore) production and atmospheric concentration. It considers the research that has investigated long-term aerobiological records as well as a range of experimental studies. Chapter 3 examines changes in the types of allergen in our environment by examining the impacts of climate change on the spatial distributions of allergenic species. It explores the evidence for range shifts of allergen-producing plant species as well as a number of stinging insects including wasps, hornets, ants, and bees. Our exposure to environmental aeroallergens also depends on the dispersion and transport of them within the atmosphere and deposition of them from the atmosphere, and it is in Chapter 4 that the impacts of climate change on these processes are considered. With so little existing research in this area, Chapter 4 presents not only an assessment of previous research but also the results of new research by the authors. While this research focusses on Europe, the methods it uses and to some extent the results it obtains provide insights that will inform future research endeavours on this aspect of the topic in other regions of the world.

Chapters 5 and 6 assess the impacts of climate change on two other important aspects of environmental allergens – their potency and timing. Chapter 5 focusses on changes in allergenicity not only of pollen but also, to a lesser extent, of mould spores, contact allergens, and food allergens. Chapter 6 then examines changes in allergen seasonality, including changes in season start dates, end dates, and durations. Again, this chapter focusses primarily on pollen, on which most of the relevant research has focused, but also to a lesser extent on fungal spores.

While much of this book focusses on outdoor environmental allergens, allergens also occur in our indoor environment. It is therefore important that the impacts of climate change on such allergens also be explicitly examined, and this is the focus of Chapter 7. The chapter considers house dust mite, cockroach, mouse, and fungal allergens. In addition to an assessment of existing literature, the chapter presents new research by the authors on the vulnerability of parts of the United States to flooding and therefore mould growth, through an analysis of the extent of homes with basements.

In the final chapter of the book to focus on the impacts of climate change on allergens per se, Chapter 8 tackles the topic of interactions among climate change, air pollutants, and aeroallergens. Following an overview of the impacts of climate change on air pollution, including ozone and particulate matter, the chapter examines the interactions between air pollutants and pollen both within the human body and in the atmosphere.

It is in Chapter 9 that the focus of the book turns explicitly to the impacts of climate change on allergic diseases, wherein a spectrum of allergic diseases is considered. Asthma and allergic rhinitis occupy much of the coverage, but other important allergic diseases include allergic conjunctivitis, atopic dermatitis, insect sting allergy, and food allergy, all of which are discussed.

Chapter 10 provides a synthesis of the discussions in the preceding chapters – a complete picture of the impacts of climate change on allergens and allergic diseases. This final chapter also provides an overview of the basic responses to the impacts of climate change, specifically mitigation and adaptation, as well as adaptation responses specific to impacts on allergens and allergic diseases. The chapter, and the book, finishes with some words of encouragement and a call to action.

References

Agard, J., Schipper, E. L. F., Birkmann, J., *et al.* (2014). Annex II: Glossary. In: Field, C. B., Barros, V. R., Dokken, D. J., *et al.*, eds. *Climate Change 2014: Impacts, Adaptation, and Vulnerability. Part A: Global and Sectoral Aspects. Contribution of Working Group II to the Fifth Assessment Report of the Intergovernmental Panel on Climate Change*. Cambridge, UK, and New York, USA: Cambridge University Press, pp. 1757–1776.

Beggs, P. J. (2014). Impacts of climate change on allergens and allergic diseases: knowledge and highlights from two decades of research. In: Butler, C. D., ed. *Climate Change and Global Health*. Wallingford, UK, and Boston, USA: CAB International, pp. 105–113.

Beggs, P. J. (2015). Environmental allergens: from asthma to hay fever and beyond. *Current Climate Change Reports*, 1(3), 176–184.

Butler, C. D., ed. (2014). *Climate Change and Global Health*. Wallingford, UK, and Boston, USA: CAB International.

Collins, M., Knutti, R., Arblaster, J., *et al.* (2013). Long-term climate change: projections, commitments and irreversibility. In: Stocker, T. F., Qin, D., Plattner, G.-K., *et al.*, eds. *Climate Change 2013: The Physical Science Basis. Contribution of Working Group I to the Fifth Assessment Report of the Intergovernmental Panel on Climate Change*. Cambridge, UK and New York, USA: Cambridge University Press, pp. 1029–1136.

Costello, A., Abbas, M., Allen, A., *et al.* (2009). Managing the health effects of climate change: *Lancet* and University College London Institute for Global Health Commission. *The Lancet*, 373(9676), 1693–1733.

Hamaoui-Laguel, L., Vautard, R., Liu, L., *et al.* (2015). Effects of climate change and seed dispersal on airborne ragweed pollen loads in Europe. *Nature Climate Change*, 5, 766–771.

Hartmann, D. L., Klein Tank, A. M. G., Rusticucci, M., *et al.* (2013). Observations: atmosphere and surface. In: Stocker, T. F., Qin, D., Plattner, G.-K., *et al.*, eds. *Climate Change 2013: The Physical Science Basis. Contribution of Working Group I to the Fifth Assessment Report of the Intergovernmental Panel on Climate Change*. Cambridge, UK and New York, USA: Cambridge University Press, pp. 159–254.

IPCC (2013a). *Climate Change 2013: The Physical Science Basis. Contribution of Working Group I to the Fifth Assessment Report of the Intergovernmental Panel on Climate Change* (Stocker, T. F., Qin, D., Plattner, G.-K., *et al.*, eds.). Cambridge, UK and New York, USA: Cambridge University Press.

IPCC (2013b). Summary for policymakers. In: Stocker, T. F., Qin, D., Plattner, G.-K., *et al.*, eds. *Climate Change 2013: The Physical Science Basis. Contribution of Working Group I to the Fifth Assessment Report of the Intergovernmental Panel on Climate Change*. Cambridge, UK and New York, USA: Cambridge University Press, pp. 3–29.

IPCC (2013c). Annex III: Glossary. In: Stocker, T. F., Qin, D., Plattner, G.-K., *et al.*, eds. *Climate Change 2013: The Physical Science Basis. Contribution of Working Group I to the Fifth Assessment Report of the Intergovernmental Panel on Climate Change*. Cambridge, UK, and New York, USA: Cambridge University Press, pp. 1447–1465.

Kennedy, R., Smith, M. (2012). Effects of aeroallergens on human health under climate change. In: Vardoulakis, S., Heaviside, C., eds. *Health Effects of Climate Change in the UK 2012: Current Evidence, Recommendations and Research Gaps*. London, UK: Health Protection Agency, pp. 83–96.

National Oceanic and Atmospheric Administration (2015a). *Trends in Atmospheric Carbon Dioxide: Recent Global CO_2*. US Department of Commerce, National Oceanic and Atmospheric Administration, Earth System Research Laboratory, Global Monitoring Division. Available at: www.esrl.noaa.gov/gmd/ccgg/trends/global.html#global

National Oceanic and Atmospheric Administration (2015b). *Trends in Atmospheric Carbon Dioxide: Mauna Loa, Hawaii: CO_2 Data*. US Department of Commerce, National Oceanic and Atmospheric Administration, Earth System Research Laboratory, Global Monitoring Division. Available at: www.esrl.noaa.gov/gmd/ccgg/trends/index.html#mlo_data

Pawankar, R., Canonica, G. W., Holgate, S. T., Lockey, R. F., eds. (2011). *World Allergy Organization (WAO) White Book on Allergy*. Milwaukee, USA: World Allergy Organization.

Pearce, N., Aït-Khaled, N., Beasley, R., *et al.* and the ISAAC Phase Three Study Group (2007). Worldwide trends in the prevalence of asthma symptoms: phase III of the

International Study of Asthma and Allergies in Childhood (ISAAC). *Thorax*, 62(9), 758–766.

Smith, K. R., Woodward, A., Campbell-Lendrum, D., *et al.* (2014). Human health: impacts, adaptation, and co-benefits. In: Field, C. B., Barros, V. R., Dokken, D. J., *et al.*, eds. *Climate Change 2014: Impacts, Adaptation, and Vulnerability. Part A: Global and Sectoral Aspects. Contribution of Working Group II to the Fifth Assessment Report of the Intergovernmental Panel on Climate Change*. Cambridge, UK, and New York, USA: Cambridge University Press, pp. 709–754.

Trenberth, K. E. (2012). Framing the way to relate climate extremes to climate change. *Climatic Change*, 115(2), 283–290.

US EPA (Environmental Protection Agency) (2009). *EPA's Endangerment Finding: Health Effects*. Available at: www.epa.gov/climatechange/Downloads/endangerment/EndangermentFinding_Health.pdf.

US EPA (2016). *Endangerment and Cause or Contribute Findings for Greenhouse Gases under Section 202(a) of the Clean Air Act*. Available at: www3.epa.gov/climatechange/endangerment/

World Health Organization (2015). *Asthma. Media Centre Fact Sheet (Number 307)*. Available at: www.who.int/mediacentre/factsheets/fs307/en/

Ziska, L., Knowlton, K., Rogers, C., *et al.* (2011). Recent warming by latitude associated with increased length of ragweed pollen season in central North America. *Proceedings of the National Academy of Sciences of the United States of America*, 108(10), 4248–4251.

2

Impacts of Climate Change on Aeroallergen Production and Atmospheric Concentration

ANNETTE MENZEL[1] AND SUSANNE JOCHNER[2]

[1]*Ecoclimatology*
Technische Universität München
[2]*Physical Geography / Landscape Ecology and Sustainable Ecosystem Development*
Catholic University of Eichstätt-Ingolstadt

2.1 Introduction

Pollen, in general, represents the central vector of gene flow among plant populations and is the major factor in reproductive success and fitness in, for example, forest communities (LaDeau and Clark, 2006). Allergenic pollen, in particular, and fungal spores constitute an important human health issue (Beggs, 2004; Huynen *et al.*, 2003). Climate change–related effects have already been observed in airborne pollen concentration and pollen production, plant (Chapter 3) and pollen distribution (Chapter 4), pollen allergenicity (Chapter 5), and timing and duration of the pollen season (Chapter 6).

Many factors have been discussed that may contribute not only to more frequent and severe allergic respiratory disease but also to new allergen sensitisation and increases in the development of allergic diseases (Chapters 3–9). One factor may be the observed increase in airborne quantities of allergenic pollen (Ziello *et al.*, 2012).

In the light of recent climate change, several plant characteristics such as plant biomass and pollen production are considered to change (e.g., Albertine *et al.*, 2014; Rogers *et al.*, 2006; Ziska *et al.*, 2003), thus affecting atmospheric pollen concentration. However, the actual concentration of airborne pollen is altered by land use/land cover changes, abundance of invasive species or disturbance (see Chapter 3), and modified by various aerobiological processes, for example, emission, dispersion/transport, and deposition – factors which are predominantly controlled by atmospheric dynamics (Dahl *et al.*, 2013; see also Chapter 4). Thus, annual sums of daily average airborne pollen concentrations (also called the annual pollen index, API) can differ considerably from what would be expected from effective pollen production and release (Frei and Gassner, 2008a). Nevertheless, the API obtained from pollen traps is believed to be an appropriate quantitative measure of the intensity of the airborne pollen season (Galán *et al.*, 2008).

The next two sections will provide information about the history of aerobiological networks and the results obtained from their long-term data. The effects of meteorological factors on pollen emission, dispersion/transport, and deposition are also briefly considered (see also Chapter 4). The results of experimental studies (Section 2.3.2) help to identify and disentangle the impacts of climate change on pollen production. Separate sections are dedicated to fungal spores and, finally, to research gaps which should encourage further work in the field of aeroallergen production and atmospheric concentration.

2.2 A Short History of Aerobiological Networks

Whereas flowering onset dates of allergenic plants have been tracked for centuries using phenological observations (Chapter 6; Menzel, 2013), research on the atmospheric concentration of pollen and mould spores is fairly recent. Early pioneers in aerobiology such as Louis Pasteur (1822–1895) or Charles H. Blackley (1820–1900) already used self-made samplers (e.g., specifically designed glass bottles) to investigate airborne pollen and mould spores (Scheifinger *et al.*, 2013).

New techniques which facilitated continuous pollen sampling emerged in the middle of the twentieth century: the Durham gravity sampler was developed in 1946 (Durham, 1946); the nowadays frequently used Hirst sampler with a suction pump in 1952 (Hirst, 1952); and the Rotoslide, a rotary impact sampler, in 1967 (Ogden and Raynor, 1967).

The first national network was initiated in 1928 by Oren C. Durham and was aimed at monitoring ragweed pollen in the United States. The network soon expanded to more than fifty stations across the country recording various types of pollen – and gradually other stations were set up in Canada, Mexico, and Cuba (Scheifinger *et al.*, 2013). A range of national European networks was founded in the late 1960s and 1970s. Among the first stations were London (1961) and Derby in the United Kingdom (1968) (Emberlin *et al.*, 1993b; Spieksma *et al.*, 2003). By 1987, shortly after its foundation, the European Aeroallergen Network had twenty-one member states with 251 sampling sites (Nilsson, 1988). It now includes more than 600 stations across Europe (https://ean.polleninfo.eu/Ean/). However, there are no or only a few long-term aerobiological records from many other parts in the world, in particular Africa, South America, and Asia.

2.3 Impacts of Climate Change

2.3.1 Long-Term Aerobiological Records

The objectives of measuring airborne pollen concentrations on a long-term basis are inter alia to evaluate the presence of different pollen types across seasons for the compilation of pollen calendars (Docampo *et al.*, 2007; Ong *et al.*, 1995), to relate the actual pollen concentration with prevailing symptoms (Frei and Gassner, 2008b; Frenz, 2001), to predict airborne pollen concentrations (Smith and Emberlin, 2005), and to study the effects of environmental parameters/climate change on the API (Frei, 1998; Vázquez *et al.*, 2003; Ziello *et al.*, 2012).

2.3.1.1 Trends in API

Most of the studies based on long-term pollen trap data revealed increasing pollen concentrations over time. Spieksma *et al.* (1995), for example, studied annual sums of daily average birch pollen concentrations in five European cities (Basel, Vienna, London, Leiden, and Stockholm) during the period 1961–1993 and found modest rising trends for all analysed stations with three cases being significant. Frei (1998) proposed a link between aerobiological data of Switzerland and climate change, but he did not incorporate meteorological data to support this suggestion. The author found increases in the API which were most pronounced for hazel (+4.6%) and birch (+3.8%) and only slightly lower for grass (+2.6%) over the period 1969–1996.

A doubling of the atmospheric pollen concentration per decade was reported by Damialis *et al.* (2007) for twelve out of sixteen species in Thessaloniki, Greece. The most pronounced increases were shown for *Platanus*, *Plantago*, and *Carpinus*. The authors suggested that the factor most likely to be responsible for these changes was increasing air temperature.

Numerous other studies have also documented an increase in API (e.g., Frei and Leuschner, 2000; Jäger *et al.*, 1996; Levetin, 1998; Rasmussen, 2002; Spieksma *et al.*, 2003; Teranishi *et al.*, 2000). In addition, a few studies (Bortenschlager and Bortenschlager, 2005; Damialis *et al.*, 2007; Frei and Gassner, 2008a) also reported an increase of peak concentrations of airborne pollen. However, some studies reported that API did not significantly increase over time. Clot (2003) analysed pollen time series of twenty-five taxa from 1979 to 1999 in Neuchâtel, Switzerland. He found significantly increasing trends only for *Alnus*, *Ambrosia*, *Artemisia*, and *Taxus*/Cupressaceae.

Other examples of plants or sites where pollen concentrations did not increase were reported by, for example, Corden and Millington (1999), Frei and Leuschner (2000), and Frei and Gassner (2008b). Non-significant trends were particularly found for grass pollen (e.g., Clot, 2003; Spieksma *et al.*, 2003). It appears to be difficult to detect trends in grass pollen concentrations since this type of pollen can only be identified at the family level. Consequently, data includes pollen from a larger number of species with overlapping flowering periods from spring to the end of summer (Spieksma *et al.*, 2003).

Most of the studies only highlighted findings from particular stations. The first comprehensive analysis of temporal changes in pollen counts was presented by Ziello *et al.* (2012). By analysing 1,221 pollen time series at ninety-seven stations across Europe, the authors provided evidence of an increase in atmospheric pollen of different taxa during the period 1977–2009 at a continental scale. Fifty-nine per cent of the series increased in API and the increase was statistically significant for 14% of the series. Forty-one per cent decreased in API and the decrease was statistically significant for 8% of the series. Analysis by taxa showed that API increased significantly for nine (Cupressaceae, *Platanus*, *Corylus*, *Fraxinus*, *Quercus*, *Alnus*, *Betula*, *Ambrosia*, and Pinaceae) and decreased significantly for two (*Artemisia* and Chenopodiaceae) of the twenty-three taxa. In general, trends in API were more pronounced for trees than for herbs or shrubs. Analysis by country showed a (significantly) positive trend in API for eleven (five) out of thirteen countries (exceptions: Spain and The Netherlands). If significant, relationships between API and local temperatures were positive for most of the species. The relationships for *Alnus*, *Betula*, and *Corylus* were negative, probably because these species are not very abundant at warmer sites. The authors suggested that not rising temperatures but carbon dioxide (CO_2) concentrations might be the decisive factor for the observed increase in the annual sum of daily averaged airborne pollen concentrations. The importance of CO_2 was also demonstrated by Zhang *et al.* (2013) who applied a Bayesian framework to project atmospheric levels of airborne birch pollen for stations in Europe and the United States. The corresponding annual mean CO_2 concentrations as well as the API of the previous year were selected as the most significant variables to model birch pollen levels. Their results suggest that annual cumulative airborne pollen count and maximum daily pollen count from 2020 to 2100 under different Intergovernmental Panel on Climate Change (IPCC) (2007) scenarios will be 1.3 to 8.0 and 1.1 to 7.3 times higher, respectively, than the mean values for 2000. In a second paper (Zhang *et al.*, 2015), it was reported that across the contiguous United States API increased by 46.0% from 1994–2000 to 2001–2010, associated with changes in growing degree days, frost-free days, and precipitation. A thorough evaluation of the effects of CO_2 on pollen production can be facilitated by experimental studies. Findings derived from these studies are presented in Section 2.3.2.

2.3.1.2 Meteorological Influences

Pollen grains are already being formed in anthers in the year previous to flowering (Emberlin *et al.*, 1990). The influence of previous summer temperature on the intensity of pollination is expected to be high since trees produce and accumulate a huge amount of photosynthates in summer which are acquired for subsequent reproduction in spring (Cadman *et al.*, 1994).

Several studies reported relationships between meteorological variables of the year prior to flowering and API. Teranishi *et al.* (2000) demonstrated a significant relationship between mean temperatures of the previous July and API of Japanese cedar (*Cryptomeria japonica*) for urban areas in Japan during 1983–1998. Latorre (1999) also reported higher annual totals of daily arboreal pollen concentrations in Mar del Plate, Argentina, resulting from favourable conditions during the summer before flowering. Hicks *et al.* (1994) detected a significant correlation between API of birch in Finland and the thermal sum of the previous year.

In addition to temperature, Rasmussen (2002) found that precipitation of May to July of the previous year was negatively correlated with the annual sum of daily averaged airborne birch pollen concentrations. However, the author suggested that this relationship was due to the commonly negative correlation between temperature and precipitation. For grass, the reverse pattern can be observed: the higher the rainfall sum of the preceding year, the higher the API in the subsequent pollen season (e.g., Schäppi *et al.*, 1998). In regions where water availability is a limiting factor (e.g., in the Mediterranean area), the influence of precipitation before and during the pollen season on pollen production and concentrations seems to be predominantly high (Dahl *et al.*, 2013). In particular, grass species are negatively affected by drought, which impedes the germination of many of their seeds, causes reduced growth, and lowers the intensity in flowering (González Minero *et al.*, 1998).

The weather conditions of the months prior to flowering are particularly important since temperature, along with photoperiod, influences the growth and development of plant species and thus controls pollen production (Laaidi, 2001). Frei (1998) found that those years which experienced a warm winter were associated with a higher API for hazel. Grass pollen concentrations were also positively correlated with the temperature of the months preceding flowering. McLauchlan *et al.* (2011) found that increasing atmospheric pollen levels in mid-North America for *Ambrosia*, Poaceae, and arboreal pollen types were associated with increasing precipitation. The authors argued that the negative effect of precipitation on pollen transport (see Section 2.3.1.4) is less important in moisture-limited regions than the potential increase in pollen production arising from high precipitation facilitating sufficient soil moisture. Negative effects of drought on pollen were observed by Gehrig (2006), who found that extended periods of negative water balance were associated with unusually small pollen loads of *Rumex*, *Urtica*, and *Artemisia* in southern Switzerland during the heat wave in 2003.

2.3.1.3 Nutrients and Pollutants

Little is known about the role of additional factors in the modification of pollen concentration and production. It has been suggested that eutrophication induced by nitrogen may have increased API over time (Damialis *et al.*, 2007). Environmental pollution (e.g., from traffic and agricultural enterprises) leads to an accumulation of nitrogen in the soil and could explain the increasing atmospheric pollen levels at five western European monitoring stations of stinging nettle that prefers soils with a high nitrogen content (Spieksma *et al.*, 2003). However, along an urbanisation gradient (see also Section 2.3.2.3), Jochner *et al.* (2013b) found that high atmospheric nitrogen dioxide (NO_2) concentration and foliar iron concentration were associated with decreased pollen production of birch in Munich.

2.3.1.4 Aerobiological Processes

Factors triggering pollen release (with temperature being the most important one) are similar to those influencing the start date of the season (Dahl *et al.*, 2013; see Chapter 6). During pollen dispersal, precipitation can lead to a washout of pollen causing a dramatic reduction of registered counts (Latorre, 1999). A negative association between birch pollen concentrations and precipitation was also detected by Frei (1998) in Switzerland. In addition to precipitation, wind is a major factor responsible for pollen dispersal (Jochner *et al.*, 2012; Laaidi, 2001; see also Chapter 4). Other factors influencing pollen dispersion are atmospheric stability and mixing height (Rasmussen, 2002). Atmospheric pollen levels might also be affected by resuspension of pollen (Vázquez *et al.*, 2003) and long-distance pollen transport (e.g., Jochner *et al.*, 2012; Rantio-Lehtimäki, 1994; see also Chapter 4). Pollen-loaded air masses can travel long distances and may enhance the concentration of common atmospheric pollen and add pollen of species that are not established in a region to the pollen spectra. This is important for allergenic pollen (e.g., of ragweed) which probably could induce new sensitisations (Zauli *et al.*, 2006).

2.3.1.5 Masting Behaviour

Differences in the atmospheric pollen concentration can also be attributed to masting, the inherent inter-annual variation of pollen production by plant populations (Ranta *et al.*, 2008) which may also mask the influence of other factors such as temperature (Frei, 1998). Alternating patterns in API are typically observed in tree pollen of boreal and temperate trees such as *Betula*, *Alnus*, *Quercus*, but also *Olea* (Dahl *et al.*, 2013). A biennial pattern was predominantly observed for birch, the most important allergenic tree species (Spieksma *et al.*, 2003).

There are many theories explaining the causes of masting (Ranta *et al.*, 2005; Spieksma *et al.*, 2003). Within the evolutionary explanations, it is hypothesised that an intermittent large reproductive effort is essential for pollination efficiency, seed production, and survival (Kelly, 1994). It is also suggested that an inter-annual variation of leaf area influences the supply of assimilates: a year with rich foliage is related to fewer catkins and followed by a year with reverse characteristics (Dahl and Strandhede, 1996). Hicks *et al.* (1994) proposed that a year with high pollen production induces high fruit production that will require much energy and restrict pollen production in the following year. Isagi *et al.* (1997) proposed that masting can be initiated by the resource balance of each plant even without any inter-annual environmental fluctuations and may result in evolutionary benefits to each individual. Conversely, meteorological conditions (sunshine, summer temperature) are supposed to lead to synchronism in the masting behaviour of plant populations (Dahl and Strandhede, 1996; Kelly, 1994; Ranta *et al.*, 2005; Schauber *et al.*, 2002). This might especially account for areas where environmental conditions represent limiting factors for reproduction (Dahl *et al.*, 2013). Furthermore, it was also found that subsequent years with favourable weather conditions even masked the biannual rhythm of birch pollen production (Dahl and Strandhede, 1996).

2.3.1.6 Other Non-meteorological Factors

Land use changes including urbanisation and the impacts of climate change on vegetation composition (see also Chapter 3) may also affect local airborne pollen concentrations. Ziello *et al.* (2012) suggested that API of some tree taxa, such as Cupressaceae which are extensively used as ornamental plants in urban areas, may have increased due to urban planning. Those authors also suggested that one aspect of land use change – afforestation – may be too slow to explain the increase in API of tree species. Grass pollen, however, is noted to be particularly affected by changes in land use (Emberlin, 1994; Frei and Gassner, 2008b). Other confounding factors may include weed control and less agricultural land being set aside (due to increasing bioenergy demand), which could have caused the decrease in taxa such as Chenopodiaceae and *Artemisia* (Ziello *et al.*, 2012). García-Mozo *et al.* (2014) analysed the olive flowering season (1982–2011) in Cordoba (Spain) and reported an increase in API partially linked to warming but also to increased olive cultivation area.

Further spread of invasive species (see Chapter 3) may also increase pollen concentrations of specific plants. For example, Clot (2003) reported an increase of ragweed pollen in Switzerland due to the expansion of this allergenic plant in France and subsequent long-range transport.

Local factors can be decisive for increases in pollen concentrations. Spieksma *et al.* (2003) found increased mugwort pollen in years when construction work was carried out next to a pollen trap in Leiden, The Netherlands. The authors explained this behaviour by a temporary increase of mugwort, which is a pioneer plant on disturbed soils.

Moreover, damage by natural disturbances or diseases may temporarily reduce pollen concentrations. Clot (2003) reported a decrease in *Ulmus* pollen related to Dutch elm disease that caused the death of many elm trees in the end of the 1980s.

2.3.2 Experimental Studies

2.3.2.1 CO_2 Enrichment and Length of the Growing Season

Plants belonging to the C_3-type photosynthesis are carbon-limited. Thus, further increases in CO_2, which is the primary source needed for photosynthesis, may result in stimulated plant growth and increased pollen production (Singer *et al.*, 2005; Ziska and Beggs, 2012).

Experimental studies in growth chambers or glasshouses under elevated CO_2 provide not only insights into future conditions but also evidence that past CO_2 increases have already resulted in significant changes in pollen production.

Wayne *et al.* (2002) used an experimental setup with a doubled CO_2 concentration (from 350 to 700 parts per million, ppm) and found a 61% increase in common ragweed pollen production. Even higher concentrations of ragweed pollen were detected by Ziska and Caulfield (2000): an increase of CO_2 from 280 (preindustrial) to 370 ppm (current level) yielded an increase of 132%; an increase from 370 to 600 ppm increased pollen production by 90%.

Differences in pollen production of timothy grass (*Phleum pratense*) under different levels of CO_2 (400 and 800 ppm) and ozone (O_3) (30 and 80 parts per billion, ppb) were investigated in a chamber experiment by Albertine *et al.* (2014). The authors found no effects of elevated O_3 but an increase of grass pollen production by ~53% per flower under higher CO_2 levels (also regardless of the O_3 treatment).

Rogers *et al.* (2006) analysed the interaction between earlier spring onset and elevated CO_2 using three different cohorts of common ragweed. The plants were exposed to ambient (380 ppm) and elevated (700 ppm) CO_2 in a climate-controlled glasshouse experiment. To simulate variability in the onset of the growing season, the germination date of the first and the second cohorts was 30 and 15 days earlier, respectively, than the last cohort. Under elevated CO_2 conditions, pollen production of the first cohort did not differ from the production of the non-fumigated equivalent, but pollen production increased by 32% in the second cohort and 55% in the last cohort. The plants that were exposed to ambient CO_2 conditions and released from dormancy earlier also experienced an increase in the production of pollen by almost 55% compared to the plants that were released 1 month later.

Thus, an advancement of phenological events may result, at least for common ragweed, in greater pollen production – regardless of actual CO_2 concentrations. This is also in agreement with long-term aerobiological studies on atmospheric pollen concentration: Teranishi *et al.* (2000) reported that a longer duration of the pollen season of Japanese cedar was associated with a higher API in urban areas of Japan. Additionally, Emberlin *et al.* (1993a) found that the duration of the grass pollen season was positively correlated with the severity of the pollen season in London, probably because both aspects depend on common variables. However,

García-Mozo *et al.* (1999) did not find a relationship between API of *Quercus* and the length of the pollen season in Spain. This is in agreement with Damialis *et al.* (2007) who reported that increasing trends in API could not be attributed to longer seasons but rather to elevated temperatures. However, since pollen production of individual trees cannot be easily translated to atmospheric pollen concentrations due to other influential factors (see Sections 2.3.1.2 to 2.3.1.6), the results of these different types of studies should be interpreted with caution.

In addition to glasshouse experiments and indoor and outdoor growth chambers, field experiments such as FACE (Free-Air CO_2 Enrichment) provide information about future conditions using CO_2-enriched air dispersed by tubes in natural settings. LaDeau and Clark (2006) utilised such a fully controlled, multiple-year experiment in the Duke Forest (North Carolina, United States) with six circular plots (30 m diameter) surrounded by thirty-two vertical pipes delivering CO_2 at multiple heights during the day. In fumigated stands (ambient CO_2+ 200 $\mu l\ l^{-1}$), the authors found significantly greater pollen cone abundance of loblolly pine (*Pinus taeda* L.) and greater pollen production. The number of pollen grains per cone and the cone length, however, were not affected by elevated CO_2.

Experiments with a more allergenic species, paper birch (*Betula papyrifera*), were conducted by Darbah *et al.* (2008) at the Aspen FACE site in Rhinelander, Wisconsin, United States. Here, the trees were exposed not only to ambient (360 ppm) and elevated (560 ppm) CO_2 but also to elevated O_3 (1.5 times ambient), as well as to elevated CO_2 plus elevated O_3 levels. Under elevated CO_2, the authors detected an increase of 140% and 70% in the total number of trees producing male flowers in the study years 2006 and 2007, respectively. The increase in the number of flowers was 260% and 100%, respectively. In contrast, elevated O_3 did not increase flowers per tree but did increase the number of trees producing male flowers. In combination, the elevated levels of the two gases only increased flower production in 2007. Catkin size was increased by 10% within the FACE experiment under elevated CO_2 conditions, but decreased by 20% under elevated O_3. The joint fumigation of these gases produced no differences in catkin length (Darbah *et al.*, 2008). In order to ensure reproductive success, carbon allocation is supposed to be enhanced under stressful conditions (e.g., under high O_3 exposure; Saikkonen *et al.*, 1998). Therefore, O_3 could be a factor in increasing pollen production, as also observed by Chappelka (2002) (but see Albertine *et al.*, 2014).

2.3.2.2 *Warming Experiments*

Wan *et al.* (2002) analysed *Ambrosia psilostachya* DC. (western ragweed) in Oklahoma, United States. They applied warming experiments with infrared heaters, increasing the air and soil temperatures by up to 1.2°C and 2.7°C, respectively. The altered environmental conditions yielded an increased number of stems and therefore also an increase in total pollen production by 84%. In conjunction with clipping, which should simulate mowing for hay, the number of stems increased by 88%. In addition, pollen diameter was significantly increased by 13%, from 21.2 to 23.9 μm. Wan *et al.* (2002) suggested that larger pollen grains might carry more allergenic compounds, but on the other hand they would sink faster and travel shorter distances than smaller pollen grains. The authors proposed a number of explanations for the increase in ragweed biomass. Its high photosynthetic rate may be advantageous in a warmer climate. Warming may prolong the length of the growing season as well as increase nitrogen availability and the release of secondary metabolites that inhibit other species. Moreover, increased light levels resulting from clipping may stimulate germination of ragweed.

2.3.2.3 Environmental Gradients

Urban–Rural Gradients/Comparisons Urban areas are of particular interest in allergological research since several studies have demonstrated that urban dwellers (especially children) are predominately influenced by pollinosis (e.g., Bibi *et al.*, 2002; Riedler *et al.*, 2000), most likely due to a westernised lifestyle, urbanisation-induced temperature increase, and abundance of air pollutants (Braun-Fahrländer *et al.*, 1999; D'Amato, 2000; Ring *et al.*, 2001; see also Chapters 8 and 9). In addition, urban areas also offer an outdoor laboratory for future pollen-related problems (Ziska *et al.*, 2003).

Interpreting data from single pollen traps located on the roofs of buildings is accompanied by difficulties since medium- and long-range transport can contribute to much higher pollen concentrations. These limitations were mentioned by Monn *et al.* (1999) who did not find any urban–rural differences in pollen counts of birch and grass species in different regions of Switzerland. In contrast to this study, Ziello *et al.* (2012) found that the increase in API across Europe was more pronounced in urban areas compared to rural and semirural areas.

Detailed information was obtained by Ziska *et al.* (2003) who arranged a number of pollen traps at a height of 1.5 m to evaluate pollen concentrations of common ragweed, planted in pots, along an urban–rural gradient in Baltimore, United States. The authors found significant increases in catkin length and plant biomass as well as higher atmospheric pollen concentrations (up to +429%) at urban locations, probably associated with increased CO_2 concentrations and air temperatures – factors which are related to urbanisation and climate change.

In general, the representativeness of pollen traps in urban areas is restricted due to the heterogeneous conditions such as the distribution of vegetation and built-up areas that may lead to differences in pollen concentrations at vertical and horizontal distances of only a few metres (Rantio-Lehtimäki *et al.*, 1991; Rapiejko, 1995). In addition, accelerated growth of the city or land use changes and urban planning (plantation) might also influence pollen concentrations (see also Section 2.3.1.6). Thus, it might be advantageous to study individual pollen production. For example, Jochner *et al.* (2011) compared pollen production (per catkin) of thirty-six birch trees in the greater area of Munich in 2009 and found no difference between urban and rural sites. A second and more extended study in the following year (Jochner *et al.*, 2013b), however, showed that high temperature along with high atmospheric NO_2 and foliar iron concentrations significantly restricted birch pollen production.

According to Emberlin *et al.* (1999), urban air pollution induces more stress on plants that may result in decreased pollen production. This is in agreement with Guedes *et al.* (2009) who reported a decreased production of pollen from white goosefoot in a polluted urban environment (Porto, Portugal). In addition, using European data, Ziello *et al.* (2012) showed that correlations between temperature trends and trends in API were negative for *Betula* and *Carpinus*. Since birch mainly grows at lower temperatures in mid- to high latitudes, temperature increases (and urban climates) could probably limit its physiological performance, expressed by a decrease in pollen production and hence API (Ziello *et al.*, 2012). This is also in agreement with Clot (2003) who suggested that a large increase in temperature could lead to lower pollen concentrations of birch but also of spruce – since these species are associated with colder climates.

Altitudinal Gradients High mountain regions may be of particular interest for people allergic to pollen since they are supposed to have a more pollen-free environment. The atmospheric pollen concentration usually decreases with increasing altitude (Clot *et al.*, 1995; Gehrig and Peeters, 2000). However, the local abundance of plant species can cause higher pollen concentrations at elevated regions (Frei, 1997; Gehrig *et al.*, 2011; Jochner *et al.*, 2012). Analysing birch and grass pollen in the Alpine region, Jochner *et al.* (2012) found that particular wind patterns can contribute to high and medically relevant pollen concentrations even at high elevation sites such as the top of Germany's highest mountain, the Zugspitze (2,962 m a.s.l.). Higher olive pollen counts at higher altitudes were observed by Aguilera and Ruiz Valenzuela (2012). The authors suggested that this may be attributable to an intrinsic mechanism compensating for a limited pollination efficiency and a short growing period.

Mountainous areas are particularly sensitive to climate change warming; thus, the Alpine region, for example, has experienced a more pronounced temperature increase compared to the adjacent lowlands (Beniston, 2006). Bortenschlager and Bortenschlager (2005) demonstrated that climate change not only affected the pollen season characteristics of lowland ecosystems, but higher altitudes in Austria were associated with higher peak values and greater total pollen counts.

2.4 Climate Change and Fungal Spores

Several species of fungal spores are important aeroallergens since they can trigger bronchial obstruction, chronic bronchitis, asthma, hypersensitivity, pneumonitis, and aspergillosis in predisposed allergic individuals (Anderson *et al.*, 1998; Stevens *et al.*, 2000). However, the effects of climate change on the growth and reproduction of fungal spores are less well documented than for allergenic pollen (Ziska *et al.*, 2008).

In contrast to plant pollen, fungal spores can be found in almost all parts of the world except in cold regions and prevail throughout the year (except during cold periods) (Caretta, 1992; Hjelmroos, 1993; Millington and Corden, 2005; Mitakakis *et al.*, 1997). Caretta (1992) estimated that the atmosphere contains 10,000–20,000 spores m^{-3} any day, whereas spores of the genera *Alternaria* and *Cladosporium*, which live as saprophytes or parasites on many kinds of plants (Hjelmroos, 1993), are among the most common and, additionally, the most studied airborne spores. *Aspergillus* and *Penicillium* spores are also dominant fungal genera and are of special importance in the indoor environment (Caretta, 1992; Levetin *et al.*, 1995; Millington and Corden, 2005; see Chapter 7).

Precipitation and increases in soil moisture stimulate fungal growth and subsequently spore liberation (Hjelmroos, 1993). Optimum growth temperatures vary between 18°C and 28°C for *Cladosporium* (Gravesen, 1979) and between 22°C and 28°C for *Alternaria* (Hjelmroos, 1993). Fungal spore occurrence fluctuates according to land use (Corden *et al.*, 2003), vegetation type (Abdel Hameed, 2005), and weather conditions (Abdel Hameed, 2005; Corden and Millington, 2001; Hjelmroos, 1993; Mitakakis *et al.*, 1997). Occurrence increases with increasing daily mean temperature, but decreases during periods of precipitation. Corden and Millington (2001) found that monthly *Alternaria* spore concentration increased during dry weather with occasional rainfall. However, there are also other spore types, frequently hyaline, that are favoured by a wet environment and found in extremely high abundance in the air just before or after rainy periods (sometimes called 'wet weather spores'; Gregory, 1973).

Long-term evaluations of spore concentrations are rare. A study by Corden and Millington (2001) showed that *Alternaria* spore concentration increased during 1970–1998 in Derby, UK, probably related to temperature increase and harvesting periods. A longer harvesting duration and more favourable mild weather conditions in autumn and winter may therefore result in higher *Alternaria* spore concentrations (Corden and Millington, 2001).

Mitakakis *et al.* (1997) found that the seasonal pattern of *Alternaria* and *Cladosporium* spore concentrations closely follows the life cycle of local vegetation: highest counts were recorded from late summer to mid-winter, which is probably related to the death of annual grasses and leaf fall.

An experimental design was presented by Klironomos *et al.* (1997): a doubling of CO_2 concentration directly affected microbial function and resulted in a greater sporulation in common soil fungi, implying a four-fold increase in the concentration of airborne propagules, mostly spores.

2.5 Research/Knowledge Gaps

The effect of increased CO_2 on ragweed and grass is an increase of plant biomass, flowers, or pollen production (see Section 2.3.2). Although ragweed has been studied extensively (Rogers *et al.*, 2006; Wayne *et al.*, 2002; Ziska *et al.*, 2003), it is still uncertain how pollen production of other herb and especially tree species will be affected by climate change. Small plants such as ragweed or grass species are easy to study because they can be planted in pots located in glasshouses or growth chambers. Further, in the natural environment, the flowers of these small-sized species are also comparably easy to count and to harvest. This is evidently in contrast to other reproducing species such as large tree species (Levetin and Van de Water, 2008) that can only be studied in FACE experiments with high effort and costs.

In any case, it is difficult to compare the results derived from glasshouse experiments and observations in the natural environment (Rogers *et al.*, 2006). For example, it was recently reported that experimental studies underestimated (Wolkovich *et al.*, 2012) and spatial studies overestimated (urban–rural gradients; Jochner *et al.*, 2013a) temperature responses in phenology compared to long-term analyses. Whereas experimental studies permit disentanglement of the effects of single or multiple possible drivers of pollen production (e.g., CO_2 alone or in combination with temperature), long-term aerobiological records integrate the effects of interactions among a number of potential influential factors that cannot be controlled (including, for example, meteorological factors and masting behaviour). In order not to overgeneralise the results derived from a particular experiment or long-term aerobiological record, results should be only interpreted with the highest caution.

There is also a need for further studies that investigate the (combined) influence of CO_2, temperature, nutrients, and other factors such as humidity and water availability whose effects are not adequately estimated or understood so far (Jochner *et al.*, 2013b).

In addition, most aerobiological studies concentrate on major allergenic plants (such as birch and grass); however, there should also be a focus on other species whose pollen is admittedly allergenic but does not constitute a major risk to the public at large. Although the implications of pollen and fungal spores are both important for allergenic disease, the latter is less well studied (Ziska *et al.*, 2008).

Since most pollen traps are installed in urban or semiurban areas (e.g., Cariñanos *et al.*, 2002; Rodríguez-Rajo *et al.*, 2010; Ziello *et al.*, 2012), it is essential to locate new traps more frequently in rural areas where the influence of climate change is not biased by an urban heat island effect and increased urban CO_2 concentrations.

Adding more years to the current data might unveil a more detailed relationship between climate change and pollen counts (Clot, 2003). In regions that only operated pollen traps for a short period or do not operate pollen traps at all, it is essential to continue or begin pollen monitoring in order to document regional variations and future impacts of climate change on atmospheric pollen concentrations.

From a statistical point of view, records longer than 30 years may be used for profound trend analyses. However, alternating patterns in pollen counts of shorter time series might reduce the statistical significance and mask the consistency of increasing trends. Therefore, it is suggested that aerobiological data of stations with fewer trees in the surrounding area are analysed. Otherwise the local effects and the small-scale behaviour of individual trees or groups of trees are recorded (Spieksma *et al.*, 2003).

The central question that still remains is whether there is a causal link between the epidemiology of allergies and increased pollen production.

References

Abdel Hameed, A. A. (2005). Vegetation: a source of air fungal bio-contaminant. *Aerobiologia*, 21(1), 53–61.

Aguilera, F., Ruiz Valenzuela, L. (2012). Altitudinal fluctuations in the olive pollen emission: an approximation from the olive groves of the south-east Iberian Peninsula. *Aerobiologia*, 28(3), 403–411.

Albertine, J. M., Manning, W. J., DaCosta, M., *et al.* (2014). Projected carbon dioxide to increase grass pollen and allergen exposure despite higher ozone levels. *PLoS One*, 9(11), e111712.

Anderson, H. R., Ponce de Leon, A., Bland, J. M., *et al.* (1998). Air pollution, pollens, and daily admissions for asthma in London 1987–92. *Thorax*, 53(10), 842–848.

Beggs, P. J. (2004). Impacts of climate change on aeroallergens: past and future. *Clinical & Experimental Allergy*, 34(10), 1507–1513.

Beniston, M. (2006). Mountain weather and climate: a general overview and a focus on climatic change in the Alps. *Hydrobiologia*, 562(1), 3–16.

Bibi, H., Shoseyov, D., Feigenbaum, D., *et al.* (2002). Comparison of positive allergy skin tests among asthmatic children from rural and urban areas living within small geographic area. *Annals of Allergy, Asthma & Immunology*, 88(4), 416–420.

Bortenschlager, S., Bortenschlager, I. (2005). Altering airborne pollen concentrations due to the Global Warming. A comparative analysis of airborne pollen records from Innsbruck and Obergurgl (Austria) for the period 1980–2001. *Grana*, 44(3), 172–180.

Braun-Fahrländer, Ch., Gassner, M., Grize, L., *et al.* (1999). Prevalence of hay fever and allergic sensitization in farmer's children and their peers living in the same rural community. *Clinical & Experimental Allergy*, 29(1), 28–34.

Cadman, A., Dames, J., Terblanche, A. P. S. (1994). Airspora concentrations in the Vaal Triangle: monitoring and potential health effects. 1, Pollen. *South African Journal of Science*, 90(11–12), 607–610.

Caretta, G. (1992). Epidemiology of allergic disease: the fungi. *Aerobiologia*, 8(3), 439–445.

Cariñanos, P., Sánchez-Mesa, J. A., Prieto-Baena, J. C., *et al.* (2002). Pollen allergy related to the area of residence in the city of Córdoba, south-west Spain. *Journal of Environmental Monitoring*, 4(5), 734–738.

Chappelka, A. H. (2002). Reproductive development of blackberry (*Rubus cuneifolius*), as influenced by ozone. *New Phytologist*, 155(2), 249–255.

Clot, B. (2003). Trends in airborne pollen: an overview of 21 years of data in Neuchâtel (Switzerland). *Aerobiologia*, 19(3–4), 227–234.

Clot, B., Peeters, A. G., Fankhauser, A., Frei, Th. (1995). *Airborne Pollen in Switzerland 1994*. Zürich: Schweizerische Meteorologische Anstalt [Swiss Meteorological Institute].

Corden, J., Millington, W. (1999). A study of *Quercus* pollen in the Derby area, UK. *Aerobiologia*, 15(1), 29–37.

Corden, J. M., Millington, W. M. (2001). The long-term trends and seasonal variation of the aeroallergen *Alternaria* in Derby, UK. *Aerobiologia*, 17(2), 127–136.

Corden, J. M., Millington, W. M., Mullins, J. (2003). Long-term trends and regional variation in the aeroallergen *Alternaria* in Cardiff and Derby UK – are differences in climate and cereal production having an effect? *Aerobiologia*, 19(3–4), 191–199.

Dahl, Å., Galán, C., Hajkova, L., *et al.* (2013). The onset, course and intensity of the pollen season. In: Sofiev, M., Bergmann, K.-C., eds. *Allergenic Pollen. A Review of the Production, Release, Distribution and Health Impacts*. Dordrecht: Springer, pp. 29–70.

Dahl, Å., Strandhede, S.-O. (1996). Predicting the intensity of the birch pollen season. *Aerobiologia*, 12(2), 97–106.

D'Amato, G. (2000). Urban air pollution and plant-derived respiratory allergy. *Clinical & Experimental Allergy*, 30(5), 628–636.

Damialis, A., Halley, J. M., Gioulekas, D., Vokou, D. (2007). Long-term trends in atmospheric pollen levels in the city of Thessaloniki, Greece. *Atmospheric Environment*, 41(33), 7011–7021.

Darbah, J. N. T., Kubiske, M. E., Nelson, N., *et al.* (2008). Effects of decadal exposure to interacting elevated CO_2 and/or O_3 on paper birch (*Betula papyrifera*) reproduction. *Environmental Pollution*, 155(3), 446–452.

Docampo, S., Recio, M., Trigo, M. M., Melgar, M., Cabezudo, B. (2007). Risk of pollen allergy in Nerja (southern Spain): a pollen calendar. *Aerobiologia*, 23(3), 189–199.

Durham, O. C. (1946). The volumetric incidence of atmospheric allergens: IV. A proposed standard method of gravity sampling, counting, and volumetric interpolation of results. *The Journal of Allergy*, 17(2), 79–86.

Emberlin, J. (1994). The effects of patterns in climate and pollen abundance on allergy. *Allergy*, 49(s18), 15–20.

Emberlin, J., Mullins, J., Corden, J., *et al.* (1999). Regional variations in grass pollen seasons in the UK, long-term trends and forecast models. *Clinical and Experimental Allergy*, 29(3), 347–356.

Emberlin, J. C., Norris-Hill, J., Bryant, R. H. (1990). A calendar for tree pollen in London. *Grana*, 29(4), 301–309.

Emberlin, J., Savage, M., Jones, S. (1993a). Annual variations in grass pollen seasons in London 1961–1990: trends and forecast models. *Clinical and Experimental Allergy*, 23(11), 911–918.

Emberlin, J., Savage, M., Woodman, R. (1993b). Annual variations in the concentrations of *Betula* pollen in the London area, 1961–1990. *Grana*, 32(6), 359–363.

Frei, T. (1997). Pollen distribution at high elevation in Switzerland: evidence for medium range transport. *Grana*, 36(1), 34–38.

Frei, T. (1998). The effects of climate change in Switzerland 1969–1996 on airborne pollen quantities from hazel, birch and grass. *Grana*, 37(3), 172–179.

Frei, T., Gassner, E. (2008a). Climate change and its impacts on birch pollen quantities and the start of the pollen season an example from Switzerland for the period 1969–2006. *International Journal of Biometeorology*, 52(7), 667–674.

Frei, T., Gassner, E. (2008b). Trends in prevalence of allergic rhinitis and correlation with pollen counts in Switzerland. *International Journal of Biometeorology*, 52(8), 841–847.

Frei, T., Leuschner, R. M. (2000). A change from grass pollen induced allergy to tree pollen induced allergy: 30 years of pollen observation in Switzerland. *Aerobiologia*, 16(3–4), 407–416.

Frenz, D. A. (2001). Interpreting atmospheric pollen counts for use in clinical allergy: allergic symptomology. *Annals of Allergy, Asthma & Immunology*, 86(2), 150–158.

Galán, C., García-Mozo, H., Vázquez, L., *et al.* (2008). Modeling olive crop yield in Andalusia, Spain. *Agronomy Journal*, 100(1), 98–104.

García-Mozo, H., Galán, C., Cariñanos, P., *et al.* (1999). Variations in the *Quercus* sp. pollen season at selected sites in Spain. *Polen*, 10, 59–69.

García-Mozo, H., Yaezel, L., Oteros, J., Galán, C. (2014). Statistical approach to the analysis of olive long-term pollen season trends in southern Spain. *Science of the Total Environment*, 473–474, 103–109.

Gehrig, R. (2006). The influence of the hot and dry summer 2003 on the pollen season in Switzerland. *Aerobiologia*, 22(1), 27–34.

Gehrig, R., Jud, S., Schuepbach, E., Clot, B. (2011). Pollen measurements in an alpine environment – altitudinal gradients and transport. In: Clot, B., Comtois, P., Escamilla-Garcia, B., eds. *Aerobiological Monographs, Towards a Comprehensive Vision*. Volume 1. MeteoSwiss (CH) and University of Montreal (CA), Montreal, Canada, pp. 19–35.

Gehrig, R., Peeters, A. G. (2000). Pollen distribution at elevations above 1000 m in Switzerland. *Aerobiologia*, 16(1), 69–74.

González Minero, F. J., Candau, P., Tomás, C., Morales, J. (1998). Airborne grass (Poaceae) pollen in southern Spain. Results of a 10-year study (1987–96). *Allergy*, 53(3), 266–274.

Gravesen, S. (1979). Fungi as a cause of allergic disease. *Allergy*, 34(3), 135–154.

Gregory, P. H. (1973). *The Microbiology of the Atmosphere*, 2nd edn. Aylesbury: Leonard Hill Books.

Guedes, A., Ribeiro, N., Ribeiro, H., *et al.* (2009). Comparison between urban and rural pollen of *Chenopodium alba* and characterization of adhered pollutant aerosol particles. *Journal of Aerosol Science*, 40(1), 81–86.

Hicks, S., Helander, M., Heino, S. (1994). Birch pollen production, transport and deposition for the period 1984–1993 at Kevo, northernmost Finland. *Aerobiologia*, 10(2–3), 183–191.

Hirst, J. M. (1952). An automatic volumetric spore trap. *The Annals of Applied Biology*, 39(2), 257–265.

Hjelmroos, M. (1993). Relationship between airborne fungal spore presence and weather variables: Cladosporium and Alternaria. *Grana*, 32(1), 40–47.

Huynen, M., Menne, B., Behrendt, H., *et al.* (2003). *Phenology and Human Health: Allergic Disorders*. Report on a WHO meeting, Rome, Italy, 16–17 January 2003. Copenhagen: WHO Regional Office for Europe.

IPCC (2007). *Climate Change 2007: Synthesis Report. Contribution of Working Groups I, II and III to the Fourth Assessment Report of the Intergovernmental Panel on*

Climate Change [Core Writing Team, Pachauri, R. K., Reisinger, A., eds.]. Geneva, Switzerland: IPCC.

Isagi, Y., Sugimura, K., Sumida, A., Ito, H. (1997). How does masting happen and synchronize? *Journal of Theoretical Biology*, 187(2), 231–239.

Jäger, S., Nilsson, S., Berggren, B., *et al.* (1996). Trends of some airborne tree pollen in the Nordic countries and Austria, 1980–1993: a comparison between Stockholm, Trondheim, Turku and Vienna. *Grana*, 35(3), 171–178.

Jochner, S. C., Beck, I., Behrendt, H., Traidl-Hoffmann, C., Menzel, A. (2011). Effects of extreme spring temperatures on urban phenology and pollen production: a case study in Munich and Ingolstadt. *Climate Research*, 49(2), 101–112.

Jochner, S., Caffarra, A., Menzel, A. (2013a). Can spatial data substitute temporal data in phenological modelling? A survey using birch flowering. *Tree Physiology*, 33(12), 1256–1268.

Jochner, S., Höfler, J., Beck, I., *et al.* (2013b). Nutrient status: a missing factor in phonological and pollen research? *Journal of Experimental Botany*, 64(7), 2081–2092.

Jochner, S., Ziello, C., Böck, A., *et al.* (2012). Spatio-temporal investigation of flowering dates and pollen counts in the topographically complex Zugspitze area on the German-Austrian border. *Aerobiologia*, 28(4), 541–556.

Kelly, D. (1994). The evolutionary ecology of mast seeding. *Trends in Ecology and Evolution*, 9(12), 465–470.

Klironomos, J. N., Rillig, M. C., Allen, M. F., *et al.* (1997). Increased levels of airborne fungal spores in response to *Populus tremuloides* grown under elevated atmospheric CO_2. *Canadian Journal of Botany*, 75(10), 1670–1673.

Laaidi, M. (2001). Forecasting the start of the pollen season of *Poaceæ*: evaluation of some methods based on meteorological factors. *International Journal of Biometeorology*, 45(1), 1–7.

LaDeau, S. L., Clark, J. S. (2006). Pollen production by *Pinus taeda* growing in elevated atmospheric CO_2. *Functional Ecology*, 20(3), 541–547.

Latorre, F. (1999). Differences between airborne pollen and flowering phenology of urban trees with reference to production, dispersal and interannual climate variability. *Aerobiologia*, 15(2), 131–141.

Levetin, E. (1998). A long-term study of winter and early spring tree pollen in the Tulsa, Oklahoma atmosphere. *Aerobiologia*, 14(1), 21–28.

Levetin, E., Shaughnessy, R., Fisher, E., *et al.* (1995). Indoor air quality in schools: exposure to fungal allergens. *Aerobiologia*, 11(1), 27–34.

Levetin, E., Van de Water, P. (2008). Changing pollen types/concentrations/distribution in the United States: fact or fiction? *Current Allergy and Asthma Reports*, 8(5), 418–424.

McLauchlan, K. K., Barnes, C. S., Craine, J. M. (2011). Interannual variability of pollen productivity and transport in mid-North America from 1997 to 2009. *Aerobiologia*, 27(3), 181–189.

Menzel, A. (2013). Europe. In: Schwartz, M. D., ed. *Phenology: An Integrative Environmental Science*, 2nd edn. Dordrecht: Springer, pp. 53–65.

Millington, W. M., Corden, J. M. (2005). Long term trends in outdoor *Aspergillus/Penicillium* spore concentrations in Derby, UK from 1970 to 2003 and a comparative study in 1994 and 1996 with the indoor air of two local houses. *Aerobiologia*, 21(2), 105–113.

Mitakakis, T., Ong, E. K., Stevens, A., Guest, D., Knox, R. B. (1997). Incidence of *Cladosporium*, *Alternaria* and total fungal spores in the atmosphere of Melbourne (Australia) over three years. *Aerobiologia*, 13(2), 83–90.

Monn, C., Alean-Kirkpatrick, P., Künzli, N., *et al.* (1999). Air pollution, climate and pollen comparisons in urban, rural and alpine regions in Switzerland (SAPALDIA study). *Atmospheric Environment*, 33(15), 2411–2416.

Nilsson, S. (1988). Preliminary inventory of aerobiological monitoring stations in Europe. *Aerobiologia*, 4(1–2), 4–7.

Ogden, E. C., Raynor, G. S. (1967). A new sampler for airborne pollen: the rotoslide. *The Journal of Allergy*, 40(1), 1–11.

Ong, E. K., Singh, M. B., Knox, R. B. (1995). Seasonal distribution of pollen in the atmosphere of Melbourne: an airborne pollen calendar. *Aerobiologia*, 11(1), 51–55.

Ranta, H., Hokkanen, T., Linkosalo, T., *et al.* (2008). Male flowering of birch: spatial synchronization, year-to-year variation and relation of catkin numbers and airborne pollen counts. *Forest Ecology and Management*, 255(3–4), 643–650.

Ranta, H., Oksanen, A., Hokkanen, T., Bondestam, K., Heino, S. (2005). Masting by *Betula*-species; applying the resource budget model to north European data sets. *International Journal of Biometeorology*, 49(3), 146–151.

Rantio-Lehtimäki, A. (1994). Short, medium and long range transported airborne particles in viability and antigenicity analyses. *Aerobiologia*, 10(2–3), 175–181.

Rantio-Lehtimäki, A., Koivikko, A., Kupias, R., Mäkinen, Y., Pohjola, A. (1991). Significance of sampling height of airborne particles for aerobiological information. *Allergy*, 46(1), 68–76.

Rapiejko, P. (1995). Monitoring Aeroalergenów w Polsce [Pollen monitoring in Poland]. In: Spiewak, R., ed. *Pollens and Pollinosis: Current Problems*. Lublin: Institute of Agricultural Medicine, pp. 13–19.

Rasmussen, A. (2002). The effects of climate change on the birch pollen season in Denmark. *Aerobiologia*, 18(3–4), 253–265.

Riedler, J., Eder, W., Oberfeld, G., Schreuer, M. (2000). Austrian children living on a farm have less hay fever, asthma and allergic sensitization. *Clinical & Experimental Allergy*, 30(2), 194–200.

Ring, J., Krämer, U., Schäfer, T., Behrendt, H. (2001). Why are allergies increasing? *Current Opinion in Immunology*, 13(6), 701–708.

Rodríguez-Rajo, F. J., Fdez-Sevilla, D., Stach, A., Jato, V. (2010). Assessment between pollen seasons in areas with different urbanization level related to local vegetation sources and differences in allergen exposure. *Aerobiologia*, 26(1), 1–14.

Rogers, C. A., Wayne, P. M., Macklin, E. A., *et al.* (2006). Interaction of the onset of spring and elevated atmospheric CO_2 on ragweed (*Ambrosia artemisiifolia* L.) pollen production. *Environmental Health Perspectives*, 114(6), 865–869.

Saikkonen, K., Koivunen, S., Vuorisalo, T., Mutikainen, P. (1998). Interactive effects of pollination and heavy metals on resource allocation in *Potentilla anserina* L. *Ecology*, 79(5), 1620–1629.

Schäppi, G. F., Taylor, P. E., Kenrick, J., Staff, I. A., Suphioglu, C. (1998). Predicting the grass pollen count from meteorological data with regard to estimating the severity of hayfever symptoms in Melbourne (Australia). *Aerobiologia*, 14(1), 29–37.

Schauber, E. M., Kelly, D., Turchin, P., *et al.* (2002). Masting by eighteen New Zealand plant species: the role of temperature as a synchronizing cue. *Ecology*, 83(5), 1214–1225.

Scheifinger, H., Belmonte, J., Buters, J., *et al.* (2013). Monitoring, modelling and forecasting of the pollen season. In: Sofiev, M., Bergmann, K.-C., eds. *Allergenic Pollen. A Review of the Production, Release, Distribution and Health Impacts*. Dordrecht: Springer, pp. 71–126.

Singer, B. D., Ziska, L. H., Frenz, D. A., Gebhard, D. E., Straka, J. G. (2005). Increasing Amb a 1 content in common ragweed (*Ambrosia artemisiifolia*) pollen as a function of rising atmospheric CO_2 concentration. *Functional Plant Biology*, 32(7), 667–670.

Smith, M., Emberlin, J. (2005). Constructing a 7-day ahead forecast model for grass pollen at north London, United Kingdom. *Clinical and Experimental Allergy*, 35(10), 1400–1406.

Spieksma, F. T. M., Corden, J. M., Detandt, M., *et al.* (2003). Quantitative trends in annual totals of five common airborne pollen types (*Betula*, *Quercus*, Poaceae, *Urtica*, and *Artemisia*), at five pollen-monitoring stations in western Europe. *Aerobiologia*, 19(3–4), 171–184.

Spieksma, F. T. M., Emberlin, J. C., Hjelmroos, M., Jäger, S., Leuschner, R. M. (1995). Atmospheric birch (*Betula*) pollen in Europe: trends and fluctuations in annual quantities and the starting dates of the seasons. *Grana*, 34(1), 51–57.

Stevens, D. A., Kan, V. L., Judson, M. A., *et al.* (2000). Practice guidelines for diseases caused by *Aspergillus*. *Clinical Infectious Diseases*, 30(4), 696–709.

Teranishi, H., Kenda, Y., Katoh, T., *et al.* (2000). Possible role of climate change in the pollen scatter of Japanese cedar *Cryptomeria japonica* in Japan. *Climate Research*, 14(1), 65–70.

Vázquez, L. M., Galán, C., Domínguez-Vilches, E. (2003). Influence of meteorological parameters on olea pollen concentrations in Córdoba (South-western Spain). *International Journal of Biometeorology*, 48(2), 83–90.

Wan, S., Yuan, T., Bowdish, S., *et al.* (2002). Response of an allergenic species, *Ambrosia psilostachya* (Asteraceae), to experimental warming and clipping: implications for public health. *American Journal of Botany*, 89(11), 1843–1846.

Wayne, P., Foster, S., Connolly, J., Bazzaz, F., Epstein, P. (2002). Production of allergenic pollen by ragweed (*Ambrosia artemisiifolia* L.) is increased in CO_2-enriched atmospheres. *Annals of Allergy, Asthma & Immunology*, 88(3), 279–282.

Wolkovich, E. M., Cook, B. I., Allen, J. M., *et al.* (2012). Warming experiments underpredict plant phenological responses to climate change. *Nature*, 485(7399), 494–497.

Zauli, D., Tiberio, D., Grassi, A., Ballardini, G. (2006). Ragweed pollen travels long distance. *Annals of Allergy, Asthma & Immunology*, 97(1), 122–123.

Zhang, Y., Bielory, L., Mi, Z., *et al.* (2015). Allergenic pollen season variations in the past two decades under changing climate in the United States. *Global Change Biology*, 21(4), 1581–1589.

Zhang, Y., Isukapalli, S. S., Bielory, L., Georgopoulos, P. G. (2013). Bayesian analysis of climate change effects on observed and projected airborne levels of birch pollen. *Atmospheric Environment*, 68, 64–73.

Ziello, C., Sparks, T. H., Estrella, N., *et al.* (2012). Changes to airborne pollen counts across Europe. *PLoS One*, 7(4), e34076.

Ziska, L. H., Beggs, P. J. (2012). Anthropogenic climate change and allergen exposure: the role of plant biology. *The Journal of Allergy and Clinical Immunology*, 129(1), 27–32.

Ziska, L. H., Caulfield, F. A. (2000). Rising CO_2 and pollen production of common ragweed (*Ambrosia artemisiifolia*), a known allergy-inducing species: implications for public health. *Australian Journal of Plant Physiology*, 27(10), 893–898.

Ziska, L. H., Epstein, P. R., Rogers, C. A. (2008). Climate change, aerobiology, and public health in the Northeast United States. *Mitigation and Adaptation Strategies for Global Change*, 13(5–6), 607–613.

Ziska, L. H., Gebhard, D. E., Frenz, D. A., *et al.* (2003). Cities as harbingers of climate change: common ragweed, urbanization, and public health. *The Journal of Allergy and Clinical Immunology*, 111(2), 290–295.

3

Impacts of Climate Change on the Distributions of Allergenic Species

LINDA J. BEAUMONT AND DAISY E. DUURSMA

Department of Biological Sciences
Faculty of Science and Engineering
Macquarie University

3.1 Introduction

All species survive within a limited climatic range, with temperature and moisture availability critically influencing distribution, abundance, and behaviour. As climate changes, therefore, populations and species may adapt in situ to new conditions or may undergo shifts in demographic patterns and distributions (Bellard *et al.*, 2012). Indeed, palaeoecological data provide evidence that distribution shifts were a common response among plants during previous episodes of climate change, altering community composition and creating new assemblages (Willis and MacDonald, 2011). Similar shifts are expected in response to anthropogenic climate change.

While a plethora of studies have assessed climate-driven distribution shifts, few have taken the perspective of allergen-producing (AP) species and ramifications for human health. Yet distribution shifts will alter the frequency of human encounters with AP species, with consequences for health care and associated economic costs. Here, we undertake a review of recent distribution shifts among AP plant and arthropod species, briefly introduce tools used to forecast future shifts, and discuss the extent to which species distributions may reconfigure as the century progresses. Although a global review, examples from the Northern Hemisphere predominate with comparatively fewer studies having been undertaken in the Southern Hemisphere.

3.2 Recent Evidence of Range Shifts

There is clear evidence that species' ranges have moved in recent decades in response to anthropogenic climate change and drivers such as habitat disturbance or destruction, land management practices, and deliberate or accidental introductions. Indeed, shifts to the distributions of a broad range of species, from all continents and most oceans, have been attributed to climate change (Bässler *et al.*, 2013; Cabrelli *et al.*, 2015; Chen *et al.*, 2011; Comte *et al.*, 2013).

3.2.1 Climate-Driven Range Shifts among AP Plant Species

In Europe, anthropogenic climate change has facilitated range shifts among numerous AP species, including establishment of walnut (*Juglans regia**[1]) in alpine valleys in Austria (Loacker *et al.*, 2007) and advancement of mountain birch (*Betula pubescens***) to higher latitudes in Norway (Hofgaard *et al.*, 2013) and higher elevations in Sweden (Kullman, 2002; Truong *et al.*, 2007).

Warming in Scandinavia has led to greater competitiveness of European beech (*Fagus sylvatica****) over Norway spruce (*Picea abies*), which has declined due to drought and attack by insect herbivores (Bolte *et al.*, 2007). In contrast, rapid declines of beech populations in southern Spain have occurred because slight increases in precipitation have been insufficient to offset the negative effect of warmer temperature on tree growth (Jump *et al.*, 2006). At higher altitudes in Spain, warmer conditions and land alterations have led to the gradual replacement of beech by holm oak (*Quercus ilex**) (Peñuelas and Boada, 2003).

Using Forest Inventory and Analysis (FIA) data from eastern United States, Woodall *et al.* (2009) concluded that eleven of fifteen northern-distributed species were migrating poleward at a rate of ~1 km yr^{-1}, including birches (*B. allerghaniensis***, *B. papyrifera***), black ash (*Fraxinus nigra**), northern pin oak (*Q. ellipsoidalis**), northern red oak (*Q. rubra**), and American basswood (*Tilia americana***). In contrast, southern-distributed species were yet to demonstrate shifts. However, Zhu *et al.* (2012), using more complete FIA data (again from eastern United States), reported contractions at both the northern and southern range margins of 58% of species, including slippery elm (*Ulmus ruba***), northern white oak (*Q. alba**), and eastern red cedar (*Juniperus virginiana**). Only 21% demonstrated a consistent northward range shift, contradicting previous generalisations that species will undergo rapid poleward shifts as climate changes.

Other studies also document range shifts in the United States, such as increased establishment of southern pioneer species in New York state (e.g., sweet birch [*B. lenta***] and sassafras [*Sassafras albidum****]; Treyger and Nowak, 2011), and uphill shifts of canyon live oak (*Q. chrysolepis**) and white burrobush (*Ambrosia dumosa**) in southern California (Kelly and Goulden, 2008). Conversely, bur oak (*Q. macrocarpa**) has demonstrated in situ adaptation whereby sensitivity to drought has decreased, potentially due to higher carbon dioxide (CO_2)-enhanced water-use efficiency: this adaptation may delay shifts in the prairie-forest ecotone in northern-central United States (Wyckoff and Bowers, 2010).

Recent range shifts among invasive AP species have been attributed to climate change, species-specific traits, land management practices, or a combination of these. One of the most significant AP species globally is common ragweed (*Ambrosia artemisiifolia**), which is also highly invasive and responsible for substantial environmental and economic damage (Box 3.1). In recent decades, range extension of this species within Europe has accelerated due to human-mediated

Box 3.1

The genus *Ambrosia*, commonly known as ragweeds, contains at least forty species mostly endemic to southwestern United States and Mexico. Ragweeds are highly allergenic: their pollen can cause hay fever, while contact with the inflorescence can result in dermatitis. Common ragweed (*Ambrosia artemisiifolia**) is one of the most serious aeroallergens worldwide (Wopfner *et al.*, 2005), with an estimated 10% of the human population experiencing allergic reactions to it (Taramarcaz *et al.*, 2005). It is also a major invasive species, inadvertently introduced across the world. This species is common in agricultural and urban areas and can germinate over a broad range of climatic and edaphic conditions, which may contribute to invasion success (Sang *et al.*, 2011).

The spread of common ragweed in Europe has accelerated in recent decades primarily due to land management (Chauvel and Cadet, 2011) and contaminated crops, bird seed, and soil (Essl *et al.*, 2009). Recent range expansions have also been facilitated by warming (Vogl *et al.*, 2008), and there is clear evidence from North America that its pollen season has lengthened (Ziska *et al.*, 2011; see Chapter 6). The urban heat island effect combined with elevated CO_2 has led to plants growing faster, flowering earlier, and having higher pollen production compared with populations in nearby rural areas (Ziska *et al.*, 2003). Growth chamber and glasshouse experiments confirm that exposure to elevated CO_2 results in higher pollen production (Wayne *et al.*, 2002; Ziska and Caulfield, 2000) and increased allergenicity (Singer *et al.*, 2005).

In coming decades, climate change is likely to enhance the invasion success of common ragweed. Significant areas of agricultural land in Austria are currently slightly too cool for this species, but populations will likely establish with only small temperature rises (Essl *et al.*, 2009). Correlative species distribution models (SDMs) project range expansion in northern China (Qin *et al.*, 2014) and northeastern Europe, with much of Europe and Russia becoming climatically suitable by 2080 (Cunze *et al.*, 2013). A mechanistic model based on weed growth, competition, and population dynamics suggested pollen production will increase at the northern Europe range margin, although ragweed populations in Spain and southern Italy will continue to be limited by drought stress (Storkey *et al.*, 2014).

Regardless of the path of movement, the economic costs associated with preventing common ragweed invasion are likely to be substantially lower than allergy treatment costs should this species continue to spread (Taramarcaz *et al.*, 2005). Across Europe, effective management of ragweed during the period 2011–2050 is likely to cost €12 billion less than the allergy costs estimated to occur in the absence of management (Richter *et al.*, 2013).

movements (Chauvel and Cadet, 2011; Essl *et al.*, 2009) and climate change (Richter *et al.*, 2013). Similarly, although great ragweed (*A. trifida**), wormwood (*Artemisia annua**), and carelessweed (*Iva xanthifolia**) were recorded in parts of central and eastern Europe in the nineteenth century, substantial range expansion has only recently occurred (Follak *et al.*, 2013). Given that temperature across large swaths of central and eastern Europe is only slightly below the requirements of these species, it is likely that their distributions will expand further in this region as climate continues to change (Follak *et al.*, 2013).

Invasion of Johnson grass (*Sorghum halepense***) into cooler regions of Austria is currently limited but also likely to expand as temperatures increase (Follak and Essl, 2013). This species was rare in Austria until the 1970s during which range expansion began, mostly due to human actions (Follak and Essl, 2013). Climate change may have facilitated the establishment of Johnson grass in Spain, as recent increases in temperature have been sufficient for germination to take place (Cirujeda *et al.*, 2011).

The Mediterranean native, red brome (*Bromus rubens* subsp. *madritensis*†), was introduced to western North America in the nineteenth century. Rapid range expansion through this region coincided with warm phases of the Pacific Decadal Oscillation (Salo, 2005). In Japan, decline in abundance of orchard grass (*Dactylis glomerata**) along its low-latitude range margin has been associated with higher mean summer temperature (Sugiyama, 2003).

Widespread modifications of herbicide usage have also altered the prevalence and distribution of invasive AP species. In Spain, annual ryegrass (*Lolium rigidum*†), a severe allergen with prolific pollen, was so uncommon in 1976 that its invasion, it was assumed, could be prevented. However, changes in herbicide usage in the 1980s aided the expansion of this species (Cirujeda *et al.*, 2011).

From 1994 to 2009, the prevalence of weeds associated with major agronomic crops in southern United States shifted, most likely due to increased usage of glyphosate herbicides (Webster and Nichols, 2012). This led to a rise in glyphosate-resistant weed species, such as Palmer amaranth (*Amaranthus palmeri**; Webster and Nichols, 2012), which has become one of the most widespread and economically expensive weeds in southeastern United States (Ward *et al.*, 2013). This plant causes severe hay fever and is a common cause of allergies among hospital admissions in Mexico (Ortega *et al.*, 2004). Several traits suggest Palmer amaranth will benefit from climate change: it is highly tolerant of heat, with peak photosynthesis occurring between 36°C and 46°C, and, as a C_4 plant, it is able to minimise water loss in hot/dry environments (Ehleringer, 1983).

3.2.1.1 Range Shifts Driven by Interactions with Other Species

Recent range shifts and changes in abundance among some species have been driven predominantly by biotic interactions. For instance, in riparian forests of the southern Appalachians (United States), the hemlock woolly adelgid, an invasive bug, has caused high mortality of eastern hemlock (*Tsuga canadensis*). This in turn has enabled deciduous trees, such as sweet birch (*B. lenta***), to increase in abundance to an extent that they will likely dominate these forests in the future (Brantley *et al.*, 2013).

While land use changes have enabled European ash (*F. excelsior**) to undergo limited range expansion in some areas of Europe (Marigo *et al.*, 2000), other populations have suffered extensive mortality due to fungal disease. First observed in Poland in the early 1990s, the disease rapidly spread to surrounding countries and now poses a substantial threat to ash populations across central and northern Europe (Pautasso *et al.*, 2013).

3.2.1.2 Urban Environments as Drivers of Species Range Shifts

Elevated temperatures in urban environments, combined with higher CO_2 concentrations and atmospheric nitrogen deposition, can facilitate establishment, growth (Searle *et al.*, 2012), and spread of invasive species. Common ragweed has been found to grow larger, emerge from seed earlier, flower earlier, and produce more pollen in urban relative to rural areas (Ziska *et al.*, 2003). In Seoul, Korea, populations of common ragweed and Japanese hop (*Humulus japonicus***), both invasive, have higher biomass in urban areas compared with populations in surrounding suburbs. Establishment of these species in urban areas is adversely impacting human populations via allergens (Song *et al.*, 2012), particularly among children for whom the age for sensitisation has declined and sensitisation rates have increased (Kim *et al.*, 2012).

The urban heat island effect combined with artificial watering can enable plants to grow in regions that would normally be outside their climatic tolerance. Within urban areas, poor planning can lead to green spaces being planted with AP species, and ornamental plants are a major cause of pollen allergies (Cariñanos and Casares-Porcel, 2011; Staffolani *et al.*, 2011). For instance, AP species constitute 37% of plant richness across metropolitan Beijing (Mao *et al.*, 2013). In La Plata, Argentina, the three most abundant tree species (*Acer* spp.) are AP non-native species and contribute substantially to hospitalisations for asthma (Nitiu and Mallo, 2002).

3.2.2 Range Shifts among Arthropods

As with other ectotherms, temperature and moisture availability can profoundly impact arthropod development, fitness, phenology, range, and migration (Drake, 1994). As such, climate change would be expected to have direct consequences on arthropod survival and distributions (Bale *et al.*, 2002). Indeed, there is substantial evidence that arthropods have responded to anthropogenic climate change via range shifts (Bässler *et al.*, 2013; Chen *et al.*, 2011) and advances in flight dates (Altermatt, 2010; Karlsson, 2014) and spring activity (Bartomeus *et al.*, 2011).

Myriad arthropods deliver bites and stings that cause pain and can result in mild to life-threatening allergic reactions, with an estimated 57–96% of people experiencing at least one sting (Antonicelli *et al.*, 2002). Stinging or swarming behaviour may be influenced by meteorological conditions. For instance, the frequency of scorpion stings is positively correlated with temperature (Molaee *et al.*, 2014). Arthropods that sting with the highest frequency belong to three families in the insect order Hymenoptera: Vespidae (wasps, hornets, yellow jackets), Formicidae (ants), and Apidae (bees; Steen *et al.*, 2005).

3.2.2.1 Vespidae

The Vespidae consists of more than 5,000 species, with many producing a painful sting, including social species such as the European wasp or yellow jacket (*Vespula germanica*), the Asian predatory wasp (*Vesta velutina*), and paper wasps (*Polistes* spp.).

Native to Palaearctic Eurasia, the European wasp is invasive in the Americas, Australia, South Africa, and New Zealand (Beggs *et al.*, 2011). Its foraging ability is remarkably plastic, and it is closely associated with human habitation (Masciocchi and Corley, 2013) which has facilitated its geographic expansion (Baz *et al.*, 2010). The wasp was reported in Chile in 1974 (Beggs *et al.*, 2011), detected in Argentina in the 1980s, and has since expanded more than 2,000 km southward and 1,000 km northward (Masciocchi and Corley, 2013). In contrast, the abundance of these wasps in England declined during the late 1970s to early 1980s, potentially due to increased insecticide usage (Archer, 2001). Cold winter temperatures cause nests to collapse and die out, limiting abundance in its native range. Invasive populations in Australia and Chile are exposed to hotter, drier conditions, and warmer winters in these countries enable some nests to survive, which greatly increases nest density in the following season (Estay and Lima, 2010; Kasper *et al.*, 2008a, 2008b). This suggests that future warming in its native range may also result in overwintering and higher abundance.

Rapid urbanisation has influenced the distribution and abundance of social Hymenoptera elsewhere. Since the 1990s, efforts to green heavily urbanised zones of Seoul, South Korea, combined with clearing of suburban forests have increased the abundance of wasps and bees in urban areas. From 2000 to 2009, reported wasp encounters increased from 706 to 3,191, while that of bees rose from 173 to 242 (Choi *et al.*, 2012). Climate change, amplified by the urban heat island effect, likely also played a role (Choi *et al.*, 2012). Similarly, from 1992 to 2005, the number of patients in Alaska, United States, seeking medical care following Hymenoptera stings increased, with greater increases in areas further north. This pattern was associated with rising annual and winter temperatures (Demain *et al.*, 2009).

3.2.2.2 Formicidae

The fire ant (*Solenopsis invicta*) is a highly invasive, aggressive, and venomous pest, native to South America. Reactions to its sting include pustules, edemia, dermal necrosis, and rarely anaphylactic shock and death (Haight and Tschinkel, 2003). Stings from fire ants are common: in infested areas, 30–60% of people are stung each year (see reviews by Kemp *et al.*, 2000, and Xu *et al.*, 2012).

The fire ant was accidently introduced into the United States prior to 1945. Populations have since expanded across southeastern United States and northeastern Mexico at an estimated rate of 193 km yr^{-1} (Kemp *et al.*, 2000). More recently, this ant has been found in the West Indies (Wetterer, 2013), Australia, New Zealand, and Taiwan and China (Ascunce *et al.*, 2011). Ecophysiological modelling has identified much of Europe, Asia, Africa, and Australia as being at risk of invasion (Morrison *et al.*, 2004).

3.2.2.3 Apidae

The genus *Apis* contains seven honeybee species, with distributions mostly centred in south and southeast Asia. Honeybees play an important role in the pollination of plants and agricultural crops and in honey production. While usually docile, stinging bees can release alarm pheromones which may elicit an attack from other bees (Brown and Tankersley, 2011).

The most widespread honeybee, *Apis mellifera*, is native to Europe, Asia, and Africa and has numerous subspecies with substantial localised variation. Destruction of forests near cities has resulted in the bees migrating to urban areas, which provide abundant cavities and more reliable food sources compared to surrounding regions (Pereira *et al.*, 2010). This has led to more frequent encounters with humans, raising concerns over public health, particularly in regions where Africanised honeybees occur (Baum *et al.*, 2008).

African honeybee queens (*A. mellifera scutellata*) were imported to Brazil in the 1950s with the aim of boosting local honey production. Unfortunately, hybridisation with other subspecies can produce aggressive colonies, and these have spread rapidly. Colonies frequently display swarming behaviour which appears linked to climate, probably via controls on food resources. In South America, swarming occurs in warmer months (Sandes Jr *et al.*, 2009), with greater bee activity and more swarms reported for periods with higher temperature and lower rainfall (de Mello *et al.*, 2003). In Tucson, Arizona, requests for colony and swarm removals increased substantially from 14 (1994) to 1,613 (2001), with more removals occurring after wet seasons (Baum *et al.*, 2008). Similarly, the emergence of

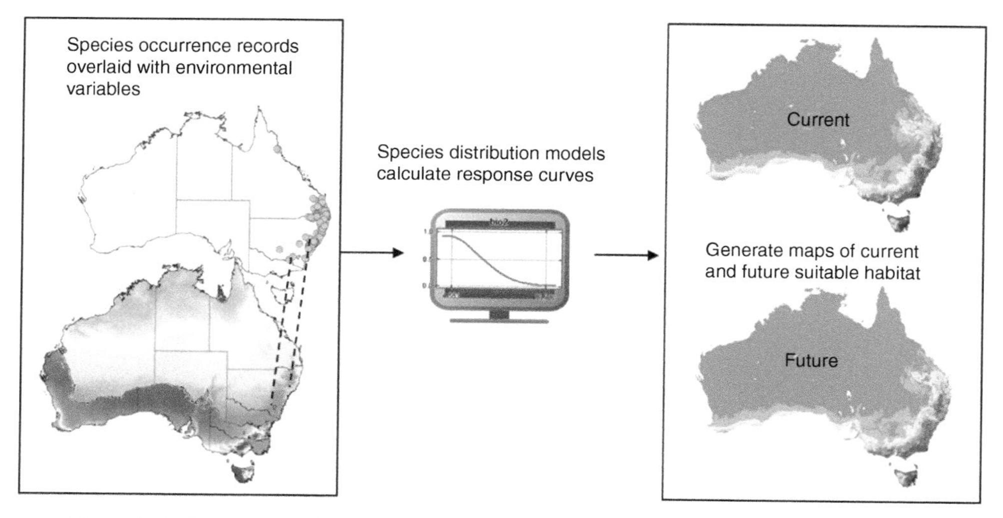

Figure 3.1. Diagrammatic representation of a correlative species distribution model. Environmental characteristics of locations where species occur are used to generate models of species–environment relationships, which can then be projected onto scenarios of climate (past, current, or future) to identify potential distributions. (A black and white version of this figure will appear in some formats. For the colour version, please refer to the plate section.)

honeybee swarms in Germany has been linked to temperature and rainfall, with swarming events increasing with the number of successive warm days (Henneken *et al.*, 2012). These patterns provide useful indications of how the frequency of honeybee–human interactions may be altered by future climate change.

3.3 Future Range Shifts among AP Species

To summarise, there is clear evidence that the distributions of a wide range of plant and animal AP species have shifted recently: there is also evidence that some shifts have increased the exposure of humans to allergens. As the twenty-first century progresses, how can we determine, a priori, if the ranges of AP species will shift further and what is the magnitude of these projected changes?

3.3.1 Tools to Assess Range Changes

Useful forecasts of biological responses to environmental change are crucial for any form of effective management, whether the aim is to maximise biodiversity or minimise negative impacts on human health. Although laboratory experiments, natural history studies, and citizen science can be used to identify biological responses to climate change, the primary tools to estimate range shifts are species distribution models (SDMs). Most assessments utilise correlative SDMs (Figure 3.1), which relate a species' occurrence to environmental conditions. The models are then used to identify geographic regions that contain suitable 'habitat' (usually defined by climate) for the target species. Projecting models onto scenarios of future climate enables maps to be created demonstrating how the distribution of suitable habitat may shift.

While correlative SDMs are process-implicit, mechanistic (or ecophysiological) models incorporate explicit relationships between environmental conditions and biological performance, which are then translated into mapped distributions. Importantly, mechanistic models can account for the 'fertilisation effect' of elevated CO_2 on plant growth, as well as biophysical thresholds, demographic patterns, and foraging energetics. However, mechanistic models are data intensive, and sufficient information for model calibration is frequently unavailable. Regardless of approach, all models are a simplification of reality and come with numerous assumptions, limitations, and sources of uncertainty (Beaumont *et al.*, 2008; Elith and Leathwick, 2009).

To date, neither correlative nor mechanistic approaches have consistently been found superior to the other (e.g., Buckley *et al.*, 2010; Dormann *et al.*, 2012), and, at times, the two approaches have produced contrasting forecasts. Keenan *et al.* (2011) modelled the potential distributions of three tree species, including holm oak, in Spain. When CO_2 was *excluded* as a predictor variable, mechanistic SDMs projected reduced productivity, due to declining climate suitability. Likewise, correlative models projected climate to become unsuitable across 40% of current stands of holm oak by 2050–2080. In contrast, incorporating CO_2 into mechanistic models resulted in increased productivity of all species.

An important assumption of SDMs is the conservation of niche; that is, environmental conditions limiting a species' distribution remain stable over space and time. Yet, among some invasive species, non-native populations may occupy climatic niches that differ from native populations (Beaumont *et al.*, 2009; Gallagher *et al.*, 2010). As a consequence, SDMs may underestimate the extent of suitable habitat.

Ultimately, although a region may be projected to contain suitable climate for the target species, a host of other factors will determine whether populations establish. Hence, SDMs should be regarded as a 'coarse filter' approach to estimating range changes.

3.3.2 *Projections of Twenty-First-Century Climate Change–Driven Shifts among Plants*

Climate change will result in significant reshuffling of plant communities. Simulations of vegetation zones across Europe show that successional shifts may lead to 31–40% of Europe being covered by a different vegetation type by 2085 (Hickler *et al.*, 2012). Generally, European trees are projected to shift northeasterly: while some species in southern Europe may benefit from range gains, those in the north may experience declines (Harrison *et al.*, 2006). Primary hotspots of change include high altitude/latitude regions, where trees are projected to replace tundra; transition zones between temperate broad-leaved and boreal conifer forest; and the conversion of forests to shrublands in southern Europe due to drought (Hickler *et al.*, 2012).

AP species in southern Europe are projected to migrate to higher elevations, leading to a decline among cold-adapted species, such as beech and juniper (*Juniperus thurifera*†; Ruiz-Labourdette *et al.*, 2012). Climate across the southern range of the arctic-alpine plant *Salix herbacea*† is projected to become unsuitable by 2080 (Alsos *et al.*, 2009), while the ranges of low-elevation species, such as holm oak (*Q. ilex* ssp. *ballota*†) and prickly cedar (*J. oxycedrus*†), which are heat- and drought-tolerant, may increase substantially (Ruiz-Labourdette *et al.*, 2012). In eastern Europe, the distributions of broad-leaved deciduous trees (e.g., *Fraxinus**, *Quercus**, and *Tilia***) are also projected to expand (Zhang *et al.*, 2009).

Populations of European ash in southern Europe are limited by hot summers, minimum winter temperature, and late spring frosts. Future precipitation declines may lead to northward range contractions, while milder winter temperature at higher latitudes may prevent chilling requirements from being met (see review by Hemery *et al.*, 2010). Poleward shifts are also projected for the birches (*B. pendula*** and *B. pubescens***). Growth of silver birch is limited by lower temperature to a greater extent than conifers, and this species may benefit disproportionably from warming, altering forest composition in Scandinavia (Hemery *et al.*, 2010).

The walnut genus contains economically and ecologically important tree species. These are suited to mild to warm climates: warmer winters in regions currently too cold for their establishment and growth, such as England and parts of Austria (Gauthier and Jacobs, 2011), may be beneficial (Hemery *et al.*, 2010). In other regions, susceptibility to drought may disadvantage members of this genus, particularly if temperature increases are not offset by higher precipitation.

The distributions of several important Mediterranean AP species are also projected to alter. Suitable habitat for hop hornbeam (*Ostrya carpinifolia*†) may expand substantially throughout central, eastern, and northern Europe (Harrison *et al.*, 2006). Among arid range species, the current distributions of mastic tree (*Pistacia lentiscus*†) and oleander (*Nerium oleander*†) are projected to remain suitable, while range expansions may occur to the northwest. Although Valonia oak (*Q. macrolepis*†) is projected to lose more than half of its current distribution in Greece, suitable habitat will remain elsewhere in Europe (Harrison *et al.*, 2006).

Archaeological and historical evidence documents the presence of olive (*Olea europeae**) cultivation in Mediterranean regions beyond this species' current geographical limits (Moriondo *et al.*, 2008). The distribution of olives is primarily limited by temperature, and contraction from southern Italy may occur due to heat stress (Gutierrez *et al.*, 2009). In southwestern United States, the southern range margin of olives may also contract, while yields of northern populations may increase. It is likely, however, that winter temperature will continue to prevent its northern range margin from expanding (Gutierrez *et al.*, 2009).

The New England region in northeastern United States is likely to be dominated by oaks by the end of this century: projections suggest that these species may cover 60% of this region compared to 21% in 2000 (Tang *et al.*, 2012). Spruce-fir and white pine-cedar forests are projected to contract to higher elevations (Tang *et al.*, 2012), and contractions may also occur among paper birch (*B. papyrifera***), aspens (*Populus grandidentata*** and *P. tremuloides***), and black cherry (*Prunus serotina****) in the northeastern United States (Iverson *et al.*, 2008). However, fragmentation of landscapes may ultimately restrict range shifts. Incorporation of colonisation potential and migration rates into models indicates that the range margins of oaks may only shift 20 km over the next 100 years – this represents a small fraction of the total area projected to be suitable (Prasad *et al.*, 2013).

Climate change may facilitate the northward expansion of numerous invasive AP species in the United States, including privet (*Ligustrum sinense**, *L. vulgare**), cogon grass (*Imperata cylindrica*†; Bradley *et al.*, 2010), and Johnson grass (McDonald *et al.*, 2009). In Australia, Mexican fireweed (*Bassia scoparia***) is classified as a 'sleeper' (Scott *et al.*, 2008); that is, although naturalised its population size has not increased exponentially (Groves, 1999). Mechanistic models suggest the risk of it establishing may increase substantially, particularly if summer rainfall increases (Scott *et al.*, 2008). *Lantana camera*†, poisonous and extremely invasive in Australia, is projected to shift into new regions in southeastern mainland Australia and Tasmania, while low-latitude areas may become unsuitable (Taylor *et al.*, 2012). The European distribution of pigweed (*Amaranthus retroflexus***) may increase up to 45%, depending on the climate scenario, while that of bitter dock (*Rumex obtusifolius***), a native European weed commonly found in gardens and arable lands, may decline 5–46% (Hyvönen *et al.*, 2012).

3.3.3 Projections of Twenty-First-Century Climate Change–Driven Shifts among Arthropods

3.3.3.1 Vespidae

The yellow-legged hornet (*Vespa velutina nigrithorax*) is widely distributed throughout temperate to subtropical regions. Accidently introduced into France in

2005, it rapidly established populations in adjoining countries. Suitable climate for this species is projected to shift northward into southern Scandinavia and eastward into Hungary, Poland, and Belarus by 2100 (Barbet-Massin *et al.*, 2013). Given its long-distance dispersal capabilities, it is unlikely that distance will prevent the hornet reaching these regions.

European wasps have several mechanisms to cope with heat: as ambient temperature rises, body temperature can be maintained by regurgitating water, while evaporative cooling keeps nests at a constant temperature (Kasper *et al.*, 2008b). However, populations in seasonally hot regions can be severely impacted if water is limited, and ambient temperature exceeding 35°C may be lethal (Kasper *et al.*, 2008b).

3.3.3.2 Formicidae

Fire ants (*Solenopsis invicta*) display several traits that enable survival in extreme conditions. Droughts may be survived by extending tunnels and chambers down to the water table (Asano and Cassill, 2012), while during floods queens and workers can link their bodies to create a 'hydrophobic surface' (i.e., a waterproof raft) until waters recede (Mlot *et al.*, 2011). Ants that survive flooding demonstrate greater defensiveness and deliver higher doses of venom compared with non-flooded ants (Papillion *et al.*, 2011).

Fire ant colony size is largest during years with higher daily temperature (Asano and Cassill, 2012) indicating that future temperature increases may facilitate population growth. Models simulating different rates of human-mediated movement of fire ants project that 10–53% of Australia may be climatically suitable for, and within reach of, this species by 2035 (Scanlan and Vanderwoude, 2006). Northern range expansions have also been projected for North America, northern Africa, and northeast China (Bertelsmeier *et al.*, 2015).

The Argentine ant (*Linepithema humile*) is limited to areas where maximum daily temperature during the hottest month ranges from 19°C to 30°C (Hartley *et al.*, 2006). Its distribution is projected to moderately decline in size (Bertelsmeier *et al.*, 2013b) as suitable habitat retracts from tropical areas and expands towards higher latitudes (Roura-Pascual *et al.*, 2004). The longevity of populations in temperate regions may also increase as temperatures rise (Cooling *et al.*, 2012). For instance, in New Zealand, populations generally survive for only 10–20 years at present, and collapse is associated with climate: while increasing temperature supports population growth, higher rainfall leads to declines (Cooling *et al.*, 2012), potentially by decreasing soil temperature (Krushelnycky *et al.*, 2005).

The Asian needle ant (*Pachycondyla chinensis*) delivers a very painful bite, and allergic reactions, including anaphylaxis, have been reported (Cho *et al.*, 2002). Much of eastern America and southeast Asia contains suitable climate for the needle ant, and its distribution is projected to increase substantially by 2080, particularly in Europe (+210%), North America (+75%), and Asia (+63%; Bertelsmeier *et al.*, 2013a). In contrast, the potential distribution of the big-headed ant (*Pheidole megacephala*) may decline with climate change (Bertelsmeier *et al.*, 2013b, 2015). However, big-headed ant populations have not yet established in many regions that are currently, and will remain, suitable. Hence, there is still considerable potential for the invasive front of this species to expand and interact with new human populations.

Invasive success may be facilitated by climate change not just through range extensions but also via increases in suitability of areas that are currently marginal (such as with Argentine ants in New Zealand). The European fire ant (*Myrmica rubra*) is one of the world's most invasive species. Although the total area of its potential distribution is projected to remain stable by 2080, relative to today, climatic suitability may increase substantially in higher latitude regions and in southern Europe (Bertelsmeier *et al.*, 2013c). This may facilitate establishment of new populations and increases in nest density.

3.4 Conclusion

Shifts in the range margins of some AP species will be an inevitable response to twenty-first-century climate change. However, unlike in previous geological eras, numerous barriers (habitat fragmentation and connectivity, poor ecosystem health, invasive species, urbanisation) exist today which may slow, or even prevent, successful migration of some species. For those unable to disperse far or fast enough, the velocity of climate change will present an additional barrier (Loarie *et al.*, 2009).

To date, few studies have explicitly addressed the implications of species range shifts in terms of human allergic diseases. Yet, clearly as range margins expand or contract, the exposure of human populations to AP plants and animals will be altered. In addition, long-distance transport of pollen (Chapter 4) may expose new groups of people living beyond the distribution of these species to pollen counts reaching clinical thresholds (Cecchi *et al.*, 2007). Future changes to atmospheric circulation patterns may alter pathways of long-distance movements, which may also increase the number of sensitised individuals (Cecchi *et al.*, 2007).

The greening of urban zones may play an important role in urban adaptation to climate change (Pramova *et al.*, 2012), and green spaces represent a key component of modern cities (Cariñanos and Casares-Porcel, 2011). However, poor planning in the selection of urban vegetation and destruction of surrounding forests has been a fundamental driver of urban allergen exposure. Future development of green zones requires clear guidelines on species selection and control of non-native AP species (Cariñanos and Casares-Porcel, 2011).

Finally, given that numerous invasive species are also allergen-producing, human health costs should be considered when assessing their financial burden. As demonstrated by Richter *et al.* (2013), the economic costs of preventing range expansions may be less than the health costs that may occur should these species spread.

Note

1 Asterisks indicate severity of pollen allergen according to PollenLibrary.com: *severe allergen; **moderate allergen; ***mild allergen; †allergenic but genus not listed in PollenLibrary.com.

References

Alsos, I. G., Alm, T., Normand, S., Brochmann, C. (2009). Past and future range shifts and loss of diversity in dwarf willow (*Salix herbacea* L.) inferred from genetics, fossils and modelling. *Global Ecology and Biogeography*, 18(2), 223–239.

Altermatt, F. (2010). Climatic warming increases voltinism in European butterflies and moths. *Proceedings of the Royal Society B: Biological Sciences*, 277(1685), 1281–1287.

Antonicelli, L., Bilò, M. B., Bonifazi, F. (2002). Epidemiology of Hymenoptera allergy. *Current Opinion in Allergy and Clinical Immunology*, 2(4), 341–346.

Archer, M. E. (2001). Changes in abundance of *Vespula germanica* and *V. vulgaris* in England. *Ecological Entomology*, 26(1), 1–7.

Asano, E., Cassill, D. L. (2012). Modeling temperature-mediated fluctuation in colony size in the fire ant, *Solenopsis invicta*. *Journal of Theoretical Biology*, 305, 70–77.

Ascunce, M. S., Yang, C.-C., Oakey, J., *et al.* (2011). Global invasion history of the fire ant *Solenopsis invicta*. *Science*, 331(6020), 1066–1068.

Bale, J. S., Masters, G. J., Hodkinson, I. D., *et al.* (2002). Herbivory in global climate change research: direct effects of rising temperature on insect herbivores. *Global Change Biology*, 8(1), 1–16.

Barbet-Massin, M., Rome, Q., Muller, F., *et al.* (2013). Climate change increases the risk of invasion by the Yellow-legged hornet. *Biological Conservation*, 157, 4–10.

Bartomeus, I., Ascher, J. S., Wagner, D., *et al.* (2011). Climate-associated phenological advances in bee pollinators and bee-pollinated plants. *Proceedings of the National Academy of Sciences of the United States of America*, 108(51), 20645–20649.

Bässler, C., Hothorn, T., Brandl, R., Müller, J. (2013). Insects overshoot the expected upslope shift caused by climate warming. *PLoS One*, 8(6), e65842.

Baum, K. A., Tchakerian, M. D., Thoenes, S. C., Coulson, R. N. (2008). Africanized honey bees in urban environments: a spatio-temporal analysis. *Landscape and Urban Planning*, 85(2), 123–132.

Baz, A., Cifrián, B., Martín-Vega, D. (2010). Distribution of the German wasp (*Vespula germanica*) and the common wasp (*Vespula vulgaris*) (Hymenoptera: Vespidae) in natural habitats in central Spain as shown by carrion-baited traps. *Sociobiology*, 55(3), 871–882.

Beaumont, L. J., Gallagher, R. V., Thuiller, W., *et al.* (2009). Different climatic envelopes among invasive populations may lead to underestimations of current and future biological invasions. *Diversity and Distributions*, 15(3), 409–420.

Beaumont, L. J., Hughes, L., Pitman, A. J. (2008). Why is the choice of future climate scenarios for species distribution modelling important? *Ecology Letters*, 11(11), 1135–1146.

Beggs, J. R., Brockerhoff, E. G., Corley, J. C., *et al.* (2011). Ecological effects and management of invasive alien Vespidae. *BioControl*, 56(4), 505–526.

Bellard, C., Bertelsmeier, C., Leadley, P., Thuiller, W., Courchamp, F. (2012). Impacts of climate change on the future of biodiversity. *Ecology Letters*, 15(4), 365–377.

Bertelsmeier, C., Guénard, B., Courchamp, F. (2013a). Climate change may boost the invasion of the Asian needle ant. *PLoS One*, 8(10), e75438.

Bertelsmeier, C., Luque, G. M., Courchamp, F. (2013b). Global warming may freeze the invasion of big-headed ants. *Biological Invasions*, 15(7), 1561–1572.

Bertelsmeier, C., Luque, G. M., Courchamp, F. (2013c). Increase in quantity and quality of suitable areas for invasive species as climate changes. *Conservation Biology*, 27(6), 1458–1467.

Bertelsmeier, C., Luque, G. M., Hoffmann, B. D., Courchamp, F. (2015). Worldwide ant invasions under climate change. *Biodiversity and Conservation*, 24(1), 117–128.

Bolte, A., Czajkowski, T., Kompa, T. (2007). The north-eastern distribution range of European beech – a review. *Forestry*, 80(4), 413–429.

Bradley, B. A., Blumenthal, D. M., Wilcove, D. S., Ziska, L. H. (2010). Predicting plant invasions in an era of global change. *Trends in Ecology and Evolution*, 25(5), 310–318.

Brantley, S., Ford, C. R., Vose, J. M. (2013). Future species composition will affect forest water use after loss of eastern hemlock from southern Appalachian forests. *Ecological Applications*, 23(4), 777–790.

Brown, T. C., Tankersley, M. S. (2011). The sting of the honeybee: an allergic perspective. *Annals of Allergy, Asthma & Immunology*, 107(6), 463–470.

Buckley, L. B., Urban, M. C., Angilletta, M. J., *et al.* (2010). Can mechanism inform species' distribution models? *Ecology Letters*, 13(8), 1041–1054.

Cabrelli, A., Beaumont, L., Hughes, L. (2015). The impacts of climate change on Australian and New Zealand flora and fauna. In: Stow, A., Maclean, N., Holwell, G. I., eds. *Austral Ark: The State of Wildlife in Australia and New Zealand.* Cambridge: Cambridge University Press, pp. 65–82.

Cariñanos, P., Casares-Porcel, M. (2011). Urban green zones and related pollen allergy: a review. Some guidelines for designing spaces with low allergy impact. *Landscape and Urban Planning*, 101(3), 205–214.

Cecchi, L., Malaspina, T. T., Albertini, R., *et al.* (2007). The contribution of long-distance transport to the presence of *Ambrosia* pollen in central northern Italy. *Aerobiologia*, 23(2), 145–151.

Chauvel, B., Cadet, É. (2011). Introduction et dispersion d'une espèce envahissante: le cas de l'ambroisie à feuilles d'armoise (*Ambrosia artemisiifolia* L.) en France. *Acta Botanica Gallica*, 158(3), 309–327.

Chen, I.-C., Hill, J. K., Ohlemüller, R., Roy, D. B., Thomas, C. D. (2011). Rapid range shifts of species associated with high levels of climate warming. *Science*, 333(6045), 1024–1026.

Cho, Y. S., Lee, Y.-M., Lee, C.-K., *et al.* (2002). Prevalence of *Pachycondyla chinensis* venom allergy in an ant-infested area in Korea. *The Journal of Allergy and Clinical Immunology*, 110(1), 54–57.

Choi, M.-B., Kim, J.-K., Lee, J.-W. (2012). Increase trend of social hymenoptera (wasps and honeybees) in urban areas, inferred from moving-out case by 119 rescue services in Seoul of South Korea. *Entomological Research*, 42(6), 308–319.

Cirujeda, A., Aibar, J., Zaragoza, C. (2011). Remarkable changes of weed species in Spanish cereal fields from 1976 to 2007. *Agronomy for Sustainable Development*, 31(4), 675–688.

Comte, L., Buisson, L., Daufresne, M., Grenouillet, G. (2013). Climate-induced changes in the distribution of freshwater fish: observed and predicted trends. *Freshwater Biology*, 58(4), 625–639.

Cooling, M., Hartley, S., Sim, D. A., Lester, P. J. (2012). The widespread collapse of an invasive species: Argentine ants (*Linepithema humile*) in New Zealand. *Biology Letters*, 8(3), 430–433.

Cunze, S., Leiblein, M. C., Tackenberg, O. (2013). Range expansion of *Ambrosia artemisiifolia* in Europe is promoted by climate change. *ISRN Ecology*, 2013, 610126.

Demain, J. G., Gessner, B. D., McLaughlin, J. B., Sikes, D. S., Foote, J. T. (2009). Increasing insect reactions in Alaska: is this related to changing climate? *Allergy and Asthma Proceedings*, 30(3), 238–243.

de Mello, M. H. S. H., da Silva, E. A., Natal, D. (2003). Abelhas africanizadas em área metropolitana do Brasil: abrigos e influências climáticas. Africanized bees in a metropolitan area of Brazil: shelters and climatic influences. *Revista de Saúde Pública*, 37(2), 237–241.

Dormann, C. F., Schymanski, S. J., Cabral, J., *et al.* (2012). Correlation and process in species distribution models: bridging a dichotomy. *Journal of Biogeography*, 39(12), 2119–2131.

Drake, V. A. (1994). The influence of weather and climate on agriculturally important insects: an Australian perspective. *Australian Journal of Agricultural Research*, 45(3), 487–509.

Ehleringer, J. (1983). Ecophysiology of *Amaranthus palmeri*, a Sonoran Desert summer annual. *Oecologia*, 57(1–2), 107–112.

Elith, J., Leathwick, J. R. (2009). Species distribution models: ecological explanation and prediction across space and time. *Annual Review of Ecology, Evolution, and Systematics*, 40, 677–697.

Essl, F., Dullinger, S., Kleinbauer, I. (2009). Changes in the spatio-temporal patterns and habitat preferences of *Ambrosia artemisiifolia* during its invasion of Austria. *Preslia*, 81(2), 119–133.

Estay, S. A., Lima, M. (2010). Combined effect of ENSO and SAM on the population dynamics of the invasive yellowjacket wasp in central Chile. *Population Ecology*, 52(2), 289–294.

Follak, S., Dullinger, S., Kleinbauer, I., Moser, D., Essl, F. (2013). Invasion dynamics of three allergenic invasive Asteraceae (*Ambrosia trifida*, *Artemisia annua*, *Iva xanthiifolia*) in central and eastern Europe. *Preslia*, 85(1), 41–61.

Follak, S., Essl, F. (2013). Spread dynamics and agricultural impact of *Sorghum halepense*, an emerging invasive species in Central Europe. *Weed Research*, 53(1), 53–60.

Gallagher, R. V., Beaumont, L. J., Hughes, L., Leishman, M. R. (2010). Evidence for climatic niche and biome shifts between native and novel ranges in plant species introduced to Australia. *Journal of Ecology*, 98(4), 790–799.

Gauthier, M.-M., Jacobs, D. F. (2011). Walnut (*Juglans* spp.) ecophysiology in response to environmental stresses and potential acclimation to climate change. *Annals of Forest Science*, 68(8), 1277–1290.

Groves, R. (1999). Sleeper weeds. In: Bishop, A. C., Boersma, M., Barnes, C. D., eds. *Proceedings of the 12th Australian Weeds Conference, 12–16 September 1999, Hobart, Tasmania*. Hobart: Tasmanian Weed Society, pp. 632–636.

Gutierrez, A. P., Ponti, L., Cossu, Q. A. (2009). Effects of climate warming on Olive and olive fly (*Bactrocera oleae* (Gmelin)) in California and Italy. *Climatic Change*, 95(1–2), 195–217.

Haight, K. L., Tschinkel, W. R. (2003). Patterns of venom synthesis and use in the fire ant, *Solenopsis invicta*. *Toxicon*, 42(6), 673–682.

Harrison, P. A., Berry, P. M., Butt, N., New, M. (2006). Modelling climate change impacts on species' distributions at the European scale: implications for conservation policy. *Environmental Science & Policy*, 9(2), 116–128.

Hartley, S., Harris, R., Lester, P. J. (2006). Quantifying uncertainty in the potential distribution of an invasive species: climate and the Argentine ant. *Ecology Letters*, 9(9), 1068–1079.

Hemery, G. E., Clark, J. R., Aldinger, E., *et al.* (2010). Growing scattered broadleaved tree species in Europe in a changing climate: a review of risks and opportunities. *Forestry*, 83(1), 65–81.

Henneken, R., Helm, S., Menzel, A. (2012). Meteorological influences on swarm emergence in honey bees (Hymenoptera: Apidae) as detected by crowdsourcing. *Environmental Entomology*, 41(6), 1462–1465.

Hickler, T., Vohland, K., Feehan, J., *et al.* (2012). Projecting the future distribution of European potential natural vegetation zones with a generalized, tree species-based dynamic vegetation model. *Global Ecology and Biogeography*, 21(1), 50–63.

Hofgaard, A., Tømmervik, H., Rees, G., Hanssen, F. (2013). Latitudinal forest advance in northernmost Norway since the early 20th century. *Journal of Biogeography*, 40(5), 938–949.

Hyvönen, T., Luoto, M., Uotila, P. (2012). Assessment of weed establishment risk in a changing European climate. *Agricultural and Food Science*, 21(4), 348–360.

Iverson, L., Prasad, A., Matthews, S. (2008). Modeling potential climate change impacts on the trees of the northeastern United States. *Mitigation and Adaptation Strategies for Global Change*, 13(5–6), 487–516.

Jump, A. S., Hunt, J. M., Peñuelas, J. (2006). Rapid climate change-related growth decline at the southern range edge of *Fagus sylvatica*. *Global Change Biology*, 12(11), 2163–2174.

Karlsson, B. (2014). Extended season for northern butterflies. *International Journal of Biometeorology*, 58(5), 691–701.

Kasper, M. L., Reeson, A. F., Austin, A. D. (2008a). Colony characteristics of *Vespula germanica* (F.) (Hymenoptera, Vespidae) in a Mediterranean climate (southern Australia). *Australian Journal of Entomology*, 47(4), 265–274.

Kasper, M. L., Reeson, A. F., Mackay, D. A., Austin, A. D. (2008b). Environmental factors influencing daily foraging activity of *Vespula germanica* (Hymenoptera, Vespidae) in Mediterranean Australia. *Insectes Sociaux*, 55(3), 288–295.

Keenan, T., Serra, J. M., Lloret, F., Ninyerola, M., Sabate, S. (2011). Predicting the future of forests in the Mediterranean under climate change, with niche- and process-based models: CO_2 matters! *Global Change Biology*, 17(1), 565–579.

Kelly, A. E., Goulden, M. L. (2008). Rapid shifts in plant distribution with recent climate change. *Proceedings of the National Academy of Sciences of the United States of America*, 105(33), 11823–11826.

Kemp, S. F., deShazo, R. D., Moffitt, J. E., Williams, D. F., Buhner II, W. A. (2000). Expanding habitat of the imported fire ant (*Solenopsis invicta*): a public health concern. *The Journal of Allergy and Clinical Immunology*, 105(4), 683–691.

Kim, J.-H., Oh, J.-W., Lee, H.-B., *et al.* (2012). Changes in sensitization rate to weed allergens in children with increased weeds pollen counts in Seoul Metropolitan Area. *Journal of Korean Medical Science*, 27(4), 350–355.

Krushelnycky, P. D., Joe, S. M., Medeiros, A. C., Daehler, C. C., Loope, L. L. (2005). The role of abiotic conditions in shaping the long-term patterns of a high-elevation Argentine ant invasion. *Diversity and Distributions*, 11(4), 319–331.

Kullman, L. (2002). Rapid recent range-margin rise of tree and shrub species in the Swedish Scandes. *Journal of Ecology*, 90(1), 68–77.

Loacker, K., Kofler, W., Pagitz, K., Oberhuber, W. (2007). Spread of walnut (*Juglans regia* L.) in an Alpine valley is correlated with climate warming. *Flora – Morphology, Distribution, Functional Ecology of Plants*, 202(1), 70–78.

Loarie, S. R., Duffy, P. B., Hamilton, H., *et al.* (2009). The velocity of climate change. *Nature*, 462(7276), 1052–1055.

Mao, Q., Ma, K., Wu, J., *et al.* (2013). Distribution pattern of allergenic plants in the Beijing metropolitan region. *Aerobiologia*, 29(2), 217–231.

Marigo, G., Peltier, J.-P., Girel, J., Pautou, G. (2000). Success in the demographic expansion of *Fraxinus excelsior* L. *Trees*, 15(1), 1–13.

Masciocchi, M., Corley, J. (2013). Distribution, dispersal and spread of the invasive social wasp (*Vespula germanica*) in Argentina. *Austral Ecology*, 38(2), 162–168.

McDonald, A., Riha, S., DiTommaso, A., DeGaetano, A. (2009). Climate change and the geography of weed damage: analysis of U.S. maize systems suggests the potential for significant range transformations. *Agriculture, Ecosystems & Environment*, 130(3–4), 131–140.

Mlot, N. J., Tovey, C. A., Hu, D. L. (2011). Fire ants self-assemble into waterproof rafts to survive floods. *Proceedings of the National Academy of Sciences of the United States of America*, 108(19), 7669–7673.

Molaee, S. M., Ahmadi, K. A., Vazirianzadeh, B., Moravvej, S. A. (2014). A climatological study of scorpion sting incidence from 2007 to 2011 in the Dezful area of Southwestern Iran, using a time series model. *Journal of Insect Science*, 14(1), 151.

Moriondo, M., Stefanini, F. M., Bindi, M. (2008). Reproduction of olive tree habitat suitability for global change impact assessment. *Ecological Modelling*, 218(1–2), 95–109.

Morrison, L. W., Porter, S. D., Daniels, E., Korzukhin, M. D. (2004). Potential global range expansion of the invasive fire ant, *Solenopsis invicta*. *Biological Invasions*, 6(2), 183–191.

Nitiu, D. S., Mallo, A. C. (2002). Incidence of allergenic pollen of *Acer* spp., *Fraxinus* spp. and *Platanus* spp. in the city of La Plata, Argentina: preliminary results. *Aerobiologia*, 18(1), 65–71.

Ortega, E. V., Vázquez, M. I. C., Tapia, J. G., Feria, A. J. M. (2004). Alergenos más frecuentes en pacientes alérgicos atendidos en un hospital de tercer nivel [Most common allergens in allergic patients admitted into a third-level hospital]. *Revista Alergia México*, 51(4), 145–150.

Papillion, A. M., Hooper-Bùi, L. M., Strecker, R. M. (2011). Flooding increases volume of venom sac in *Solenopsis invicta* (Hymenoptera: Formicidae). *Sociobiology*, 57(2), 301–308.

Pautasso, M., Aas, G., Queloz, V., Holdenrieder, O. (2013). European ash (*Fraxinus excelsior*) dieback – a conservation biology challenge. *Biological Conservation*, 158, 37–49.

Peñuelas, J., Boada, M. (2003). A global change-induced biome shift in the Montseny mountains (NE Spain). *Global Change Biology*, 9(2), 131–140.

Pereira, A. M., Chaud-Netto, J., Bueno, O. C., Arruda, V. M. (2010). Relationship among *Apis mellifera* L. stings, swarming and climate conditions in the city of Rio Claro, SP, Brazil. *The Journal of Venomous Animals and Toxins including Tropical Diseases*, 16(4), 647–653.

Pramova, E., Locatelli, B., Djoudi, H., Somorin, O. A. (2012). Forests and trees for social adaptation to climate variability and change. *Wiley Interdisciplinary Reviews: Climate Change*, 3(6), 581–596.

Prasad, A. M., Gardiner, J. D., Iverson, L. R., Matthews, S. N., Peters, M. (2013). Exploring tree species colonization potentials using a spatially explicit simulation model: implications for four oaks under climate change. *Global Change Biology*, 19(7), 2196–2208.

Qin, Z., Ditommaso, A., Wu, R. S., Huang, H. Y. (2014). Potential distribution of two *Ambrosia* species in China under projected climate change. *Weed Research*, 54(5), 520–531.

Richter, R., Berger, U. E., Dullinger, S., *et al.* (2013). Spread of invasive ragweed: climate change, management and how to reduce allergy costs. *Journal of Applied Ecology*, 50(6), 1422–1430.

Roura-Pascual, N., Suarez, A. V., Gómez, C., *et al.* (2004). Geographical potential of Argentine ants (*Linepithema humile* Mayr) in the face of global climate change. *Proceedings of the Royal Society of London B*, 271(1557), 2527–2534.

Ruiz-Labourdette, D., Nogués-Bravo, D., Ollero, H. S., Schmitz, M. F., Pineda, F. D. (2012). Forest composition in Mediterranean mountains is projected to shift along the entire elevational gradient under climate change. *Journal of Biogeography*, 39(1), 162–176.

Salo, L. F. (2005). Red brome (*Bromus rubens* subsp. *madritensis*) in North America: possible modes for early introductions, subsequent spread. *Biological Invasions*, 7(2), 165–180.

Sandes Jr, R. L., Oliveira, C. L., Ferreira, E. S., *et al.* (2009). Spatial analysis of migrating *Apis mellifera* colonies in Salvador, Bahia, Brazil. *Geospatial Health*, 4(1), 129–134.

Sang, W., Liu, X., Axmacher, J. C. (2011). Germination and emergence of *Ambrosia artemisiifolia* L. under changing environmental conditions in China. *Plant Species Biology*, 26(2), 125–133.

Scanlan, J. C., Vanderwoude, C. (2006). Modelling the potential spread of *Solenopsis invicta* Buren (Hymenoptera: Formicidae) (red imported fire ant) in Australia. *Australian Journal of Entomology*, 45(1), 1–9.

Scott, J., Batchelor, K., Ota, N., Yeoh, P. (2008). *Modelling Climate Change Impacts on Sleeper and Alert Weeds: Final Report*. Wembley, Australia: CSIRO Entomology.

Searle, S. Y., Turnbull, M. H., Boelman, N. T., *et al.* (2012). Urban environment of New York City promotes growth in northern red oak seedlings. *Tree Physiology*, 32(4), 389–400.

Singer, B. D., Ziska, L. H., Frenz, D. A., Gebhard, D. E., Straka, J. G. (2005). Increasing Amb a 1 content in common ragweed (*Ambrosia artemisiifolia*) pollen as a function of rising atmospheric CO_2 concentration. *Functional Plant Biology*, 32(7), 667–670.

Song, U., Mun, S., Ho, C.-H., Lee, E. J. (2012). Responses of two invasive plants under various microclimate conditions in the Seoul Metropolitan Region. *Environmental Management*, 49(6), 1238–1246.

Staffolani, L., Velasco-Jiménez, M. J., Galán, C., Hruska, K. (2011). Allergenicity of the ornamental urban flora: ecological and aerobiological analyses in Córdoba (Spain) and Ascoli Piceno (Italy). *Aerobiologia*, 27(3), 239–246.

Steen, C. J., Janniger, C. K., Schutzer, S. E., Schwartz, R. A. (2005). Insect sting reactions to bees, wasps, and ants. *International Journal of Dermatology*, 44(2), 91–94.

Storkey, J., Stratonovitch, P., Chapman, D. S., Vidotto, F., Semenov, M. A. (2014). A process-based approach to predicting the effect of climate change on the distribution of an invasive allergenic plant in Europe. *PLoS One*, 9(2), e88156.

Sugiyama, S. (2003). Geographical distribution and phenotypic differentiation in populations of *Dactylis glomerata* L. in Japan. *Plant Ecology*, 169(2), 295–305.

Tang, G., Beckage, B., Smith, B. (2012). The potential transient dynamics of forests in New England under historical and projected future climate change. *Climatic Change*, 114(2), 357–377.

Taramarcaz, P., Lambelet, C., Clot, B., Keimer, C., Hauser, C. (2005). Ragweed (Ambrosia) progression and its health risks: will Switzerland resist this invasion? *Swiss Medical Weekly*, 135(37–38), 538–548.

Taylor, S., Kumar, L., Reid, N. (2012). Impacts of climate change and land-use on the potential distribution of an invasive weed: a case study of *Lantana camara* in Australia. *Weed Research*, 52(5), 391–401.

Treyger, A. L., Nowak, C. A. (2011). Changes in tree sapling composition within powerline corridors appear to be consistent with climatic changes in New York State. *Global Change Biology*, 17(11), 3439–3452.

Truong, C., Palmé, A. E., Felber, F. (2007). Recent invasion of the mountain birch *Betula pubescens* ssp. *tortuosa* above the treeline due to climate change: genetic and ecological study in northern Sweden. *Journal of Evolutionary Biology*, 20(1), 369–380.

Vogl, G., Smolik, M., Stadler, L.-M., *et al.* (2008). Modelling the spread of ragweed: effects of habitat, climate change and diffusion. *The European Physical Journal Special Topics*, 161(1), 167–173.

Ward, S. M., Webster, T. M., Steckel, L. E. (2013). Palmer amaranth (*Amaranthus palmeri*): a review. *Weed Technology*, 27(1), 12–27.

Wayne, P., Foster, S., Connolly, J., Bazzaz, F., Epstein, P. (2002). Production of allergenic pollen by ragweed (*Ambrosia artemisiifolia* L.) is increased in CO_2-enriched atmospheres. *Annals of Allergy, Asthma & Immunology*, 88(3), 279–282.

Webster, T. M., Nichols, R. L. (2012). Changes in the prevalence of weed species in the major agronomic crops of the southern United States: 1994/1995 to 2008/2009. *Weed Science*, 60(2), 145–157.

Wetterer, J. K. (2013). Exotic spread of *Solenopsis invicta* Buren (Hymenoptera: Formicidae) beyond North America. *Sociobiology*, 60(1), 50–55.

Willis, K. J., MacDonald, G. M. (2011). Long-term ecological records and their relevance to climate change predictions for a warmer world. *Annual Review of Ecology, Evolution, and Systematics*, 42, 267–287.

Woodall, C. W., Oswalt, C. M., Westfall, J. A., *et al.* (2009). An indicator of tree migration in forests of the eastern United States. *Forest Ecology and Management*, 257(5), 1434–1444.

Wopfner, N., Gadermaier, G., Egger, M., *et al.* (2005). The spectrum of allergens in ragweed and mugwort pollen. *International Archives of Allergy and Immunology*, 138(4), 337–346.

Wyckoff, P. H., Bowers, R. (2010). Response of the prairie–forest border to climate change: impacts of increasing drought may be mitigated by increasing CO_2. *Journal of Ecology*, 98(1), 197–208.

Xu, Y., Huang, J., Zhou, A., Zeng, L. (2012). Prevalence of *Solenopsis invicta* (Hymenoptera: Formicidae) venom allergic reactions in mainland China. *Florida Entomologist*, 95(4), 961–965.

Zhang, F., Li, Y., Guo, Z., Murray, B. R. (2009). Climate warming and reproduction in Chinese alligators. *Animal Conservation*, 12(2), 128–137.

Zhu, K., Woodall, C. W., Clark, J. S. (2012). Failure to migrate: lack of tree range expansion in response to climate change. *Global Change Biology*, 18(3), 1042–1052.

Ziska, L. H., Caulfield, F. A. (2000). Rising CO_2 and pollen production of common ragweed (*Ambrosia artemisiifolia*), a known allergy-inducing species: implications for public health. *Australian Journal of Plant Physiology*, 27(10), 893–898.

Ziska, L. H., Gebhard, D. E., Frenz, D. A., *et al.* (2003). Cities as harbingers of climate change: common ragweed, urbanization, and public health. *The Journal of Allergy and Clinical Immunology*, 111(2), 290–295.

Ziska, L., Knowlton, K., Rogers, C., *et al.* (2011). Recent warming by latitude associated with increased length of ragweed pollen season in central North America. *Proceedings of the National Academy of Sciences of the United States of America*, 108(10), 4248–4251.

4

Impacts of Climate Change on Aeroallergen Dispersion, Transport, and Deposition

MIKHAIL SOFIEV AND MARJE PRANK
Finnish Meteorological Institute

4.1 Introduction

The current chapter considers the possible changes in atmospheric transport of aeroallergens that can be expected due to climate change. The main attention is given to pollen, for which dispersion is studied better than for mould spores, but the main conclusions are valid for mould spores too.

The analysis starts from the pollen release from the plant, thus leaving out all processes of plant development and pollen formation (Figure 4.1). These parts of the pollen life cycle, also being vulnerable to changes in climate, are considered in different chapters of this book. The targets of this chapter are the processes related to the pollen release at local scale, its multiscale transport, and removal from the atmosphere.

This chapter heavily relies on basic information on pollen behaviour in the atmosphere, which has been discussed in Chapter 5 of Sofiev and Bergmann (2013). Below, the key points from that book are summarised for the sake of completeness of the presentation. For details, the reader is referred to the main source.

4.1.1 Pollen Features and Meteorological Parameters Controlling Its Transport in the Air

There are many meteorological processes affecting pollen behaviour in the atmosphere (Di-Giovanni and Kevan, 1991; Helbig *et al.*, 2004; Linskens and Cresti, 2000; Sofiev *et al.*, 2013). From the point of view of the atmospheric transport, pollen and, to a large extent, mould spores are coarse but light aerosols with density around 800 kg m^{-3}, which makes them from 20% to five times lighter than most particles in the air. As a result, many types of bioaerosols can be transported over hundreds of kilometres or further if atmospheric conditions appear favourable. Experience with pollen dispersion modelling shows that the pollen grains of

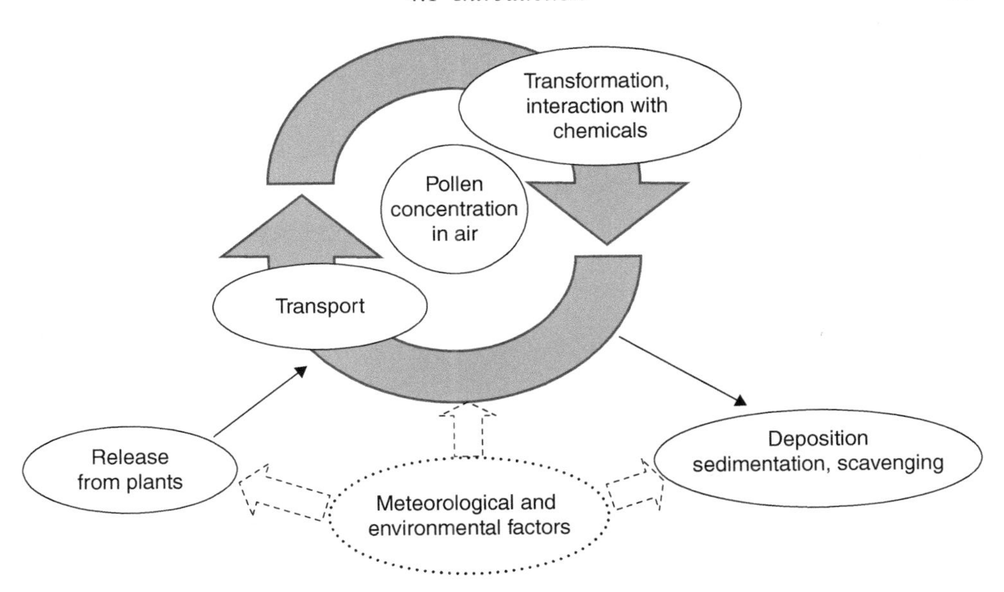

Figure 4.1. A schematic view of the main parts of pollen life cycle in the atmosphere (modified from Sofiev and Bergmann, 2013).

many wind-pollinated species, such as birch and olive trees, should be considered as regional- to continental-scale tracers. Coarser pollens (e.g., from many grasses) are important in regional to local scales.

One of the key parameters regulating pollen release to the atmosphere is the intensity of vertical mixing. It controls two processes: initial elevation of pollen from inflorescences upwards to the middle and upper parts of the boundary layer (where the main transport takes place), and remixing of particles during the transport, thus preventing them from settling down.

Large size and low density make pollen susceptible to resuspension from the ground to the air at comparatively low wind speeds and turbulence intensity.

For all pollen types, the primary dry deposition process is sedimentation. Pollen is a hygroscopic particle: in humid air it absorbs water from the air, thus increasing its own mass and/or size, whereas dry air results in reduction of water content of pollen, thus decreasing its density and/or shrinking it. The parameter controlling these processes is relative humidity. The changes in the pollen size and density, in turn, control its sedimentation velocity and lifetime in the air.

Scavenging by precipitation takes place mainly below the raining cloud via impaction mechanism; that is, the grains are hit and dragged towards the surface by the falling droplets. In-cloud processes are presumably of lower importance since pollen is primarily transported inside the boundary layer. One can still expect that pollen gets embedded into the forming droplets inside the cloud. Then the fraction of rained-out cloud water determines the fraction of pollen deposited from the cloud.

For mould spores, essentially the same mechanisms are relevant, but their smaller size makes them potentially longer lasting in the air. On the other hand, spores are often released close to the ground, which reduces the fraction reaching the main part of the boundary layer and increases the influence of vertical mixing on spore concentration at regional scale.

4.1.2 Existing Studies and Datasets, Selection of Meteorological Variables for Analysis

There is a vast variety of studies considering the ongoing and future changes of meteorological conditions, but only a fraction of them is relevant for atmospheric transport of coarse aerosols. Very few deal with pollen specifically. For instance, Prank *et al.* (2013) analysed the ragweed pollen transport changes due to meteorological factors. The SILAM dispersion model (http://silam.fmi.fi) was run through 7 years (2005–2011) with no changes in plant habitat and pollen emission parameters; that is, the difference between the years was only due to meteorology. The model predicted a steady increase of ragweed pollen concentrations in Hungary, the most infected area in central Europe. During the analysed period, the meteorology-induced growth was about 3% per year, which accounted for about half of the observed trends in that region. However, extrapolation of these findings to climate-relevant periods is hardly possible.

Since pollen-specific studies are very scarce and inconclusive, we shall consider the processes common for all coarse particles, thus benefitting from general climate studies, especially the ones assessing future aerosol concentrations.

A comprehensive overview of climate trends can be found in the report of Working Group I of the Intergovernmental Panel on Climate Change (IPCC), which collated and summarised findings of a vast amount of climate studies (IPCC, 2013). However, the specific purpose of the current assessment requires a slightly different viewpoint than what is taken by IPCC and the bulk of climate studies.

Apart from the literature studies and IPCC assessments, the analysis will also look into trends using the existing, over-30-year-long meteorological re-analysis ERA-Interim of the European Centre for Medium-Range Weather Forecasts. This dataset is homogeneous with regard to the underlying state-of-the-art meteorological model and is driven by a large variety of observational information (albeit varying from year to year) assimilated every 12 hours (Dee *et al.*, 2011). Changes in the assimilated datasets evidently affected the homogeneity of this reanalysis and posed limitations on its applicability to climate-relevant trend studies. However, the most significant variations of the assimilated data refer to the upper layers of the atmosphere. The tropospheric and surface observations are comparatively homogeneous, which is sufficient for the purposes of the current analysis.

By far the largest attention in most climate-related studies is given to mean temperature and its variability in space and time – but this is one of the least important parameters for airborne pollen transport. It strongly affects plant growth and maturation of pollen but not its dispersion. One can find correlation between temperature and, for instance, turbulent mixing and convective flows, but this is still an indirect impact. Other parameters considered below represent the atmospheric mixing much more directly.

The other heavily studied quantity is total sum of precipitation (annual or seasonal). Scavenging with precipitation is the most important removal mechanism quickly cleaning the whole depth of the boundary layer and adjacent parts of the lower troposphere. It is certainly important for pollen transport but is again rather indirect. Since pollen grains are coarse, their removal is very efficient and, as a result, even a small amount of rain completely cleans the air and the actual amount of precipitated water becomes unimportant. More relevant parameters would be the total rain duration during the pollen season representing the rainy periods with quick removal of pollen from the air and dry periods when pollen can travel undisturbed. A good representation of this characteristic – included in the IPCC analysis – is duration of dry spells.

The third parameter mentioned in connection with climate change is wind speed and, to a smaller extent, prevailing wind directions.

Relative humidity is quite rarely studied in climate applications. In the below analysis, we shall therefore make some basic statistics based on meteorological past-time data.

Parameters describing the turbulent mixing can be either boundary layer thickness (indirect but frequently used characteristic of the mixing intensity) or eddy diffusivity coefficient (directly describing the turbulence intensity but quite poorly known and verified).

To focus the analysis, the selection of the relevant atmospheric processes is based on the following requirements. First, only parameters with major impact on aerosol transport and lifetime in the atmosphere are considered. Second, only parameters directly available or easily computable from the past-time meteorological archives are included. Preference is given to the parameters directly verifiable via comparison with long-term in situ meteorological observations. Third, we consider only parameters with satisfactory quality in climate change scenarios and climate models. This particular requirement is not easy to satisfy: climate models have numerous shortcuts and simplifications, which invalidate many otherwise useful parameters.

The above considerations lead to the following list of parameters for the analysis:

1. Transport velocity can be represented by wind speed at 10 m above the ground (U_{10}), the directly observable wind characteristic that affects both pollen release and transport. It is also available from several climate-related studies and ERA-Interim.
2. Analogously, wind direction at 10 m above the ground (ϕ_{10}) can be used to describe the transport direction.
3. Fraction of rainy periods per month (τ_{pr}) can be used as the main precipitation characteristic. It is available from IPCC, numerous climate studies, and ERA-Interim.
4. For vertical mixing in the atmospheric boundary layer (ABL), we have chosen turbulent diffusion coefficient at 1 m above the ground (K_{z1}), which is the most direct parameter controlling the uplift of pollen released near the ground to the main transport layers of the air. This parameter is not available from climate studies (similarly to other parameters, such as ABL height); therefore, we shall diagnose it from ERA-Interim data.
5. The second complicated quantity is the air relative humidity. Most climate studies assume it is roughly constant and do not evaluate it. However, its near-surface value Q_2 taken at 2 m above the ground is comparatively well validated in ERA-Interim, which still makes it useful.

The analysis of the above quantities is made on a monthly basis for all seasons, but below we concentrate on just two months: April for the most important Northern Hemisphere spring-time flowering species, such as birch, olive, and some grasses, and August to reflect the conditions for Northern Hemisphere late flowering species, such as ragweed and mugwort.

All trends are analysed for three sub-regions of Europe: longitude–latitude rectangles selected in Spain (6°W–3°W, 38°N–41°N), Germany (10°E–13°E, 49°N–52°N), and Finland (22°E–25°E, 61°N–64°N) and labelled as southern, central, and northern sub-domains, respectively.

4.2 Climate-Induced Changes of the Key Meteorological Factors

This section is dedicated to current and projected future trends in the above five primary meteorological factors controlling the aerosol transport.

4.2.1 Changes in Large-Scale Atmospheric Circulations

Climate change will lead to certain modifications of the large-scale atmospheric flows. Changes in the circulations will be the large-scale drivers that control the mean wind speed and direction, as well as humidity and precipitation patterns, also affecting the scales relevant for pollen transport.

For Europe, probably the most important change will be the poleward expansion of the Hadley cells (two nearly closed circulations of hot air rising at equator and sinking at about 30°N and 30°S). Climate models project their expansion of about a quarter of a degree as a mean for the period 2016–2035 in comparison with the mean for the period 1986–2005. This would cause hotter weather in southern Europe, but also reduction of the zonal (west to east) wind in the Mediterranean latitudes with simultaneous increase in central Europe. However, the IPCC, quoting high complexity of the expansion mechanism and uncertainties in future concentrations of the key atmospheric pollutants, has put low confidence in the quantitative projections of the changes (Kirtman *et al.*, 2013).

4.2.2 Wind Speed

The impact of this parameter on pollen transport is two-fold. First, faster wind speed implies longer distance travelled by the particles before they settle down or get scavenged out. Second, many plants need mechanical forcing for pollen take-off from inflorescences. There is also an indirect connection to boundary layer turbulence: near-surface wind speed generates turbulent eddies, thus promoting the vertical mixing.

Impact on wind speed from the ongoing climate change is small and mainly negative (Horton *et al.*, 2014; McVicar *et al.*, 2012). As shown by 30-year-long analysis of observation time series, the global mean trend is -0.0148 m s^{-1} yr^{-1} of the near-surface wind speed (McVicar *et al.*, 2012). In the same work, the European observations showed significantly smaller trends – but a large fraction of eastern and northern Europe did not have sufficiently long homogeneous observations to meet the requirements of the study. The Kirtman *et al.* (2013) ensemble of forty-one global models showed future zonal (west to east) wind reduction trends over much of Europe too: over 30 years the mean wind speed at 850 hPa (height ~2 km) is projected to grow by ~0.1 m s^{-1} in central Europe and decrease by 0.1–0.2 m s^{-1} in southern and northern parts. A global study by Liao *et al.* (2006) also showed very moderate but quite homogeneous reduction of near-surface wind speed – by <0.2 m s^{-1} by 2100 over most of Europe.

The outcome of these studies is supported by the ERA-Interim dataset, which has assimilated routine meteorological observations (Figure 4.2): some negative trends are noticeable only in central region in April and the northern region in August, but their statistical significance is low (p ~0.1–0.15). Intriguingly, summer in the south is getting slightly windier – but in the last 10 years, variability has increased.

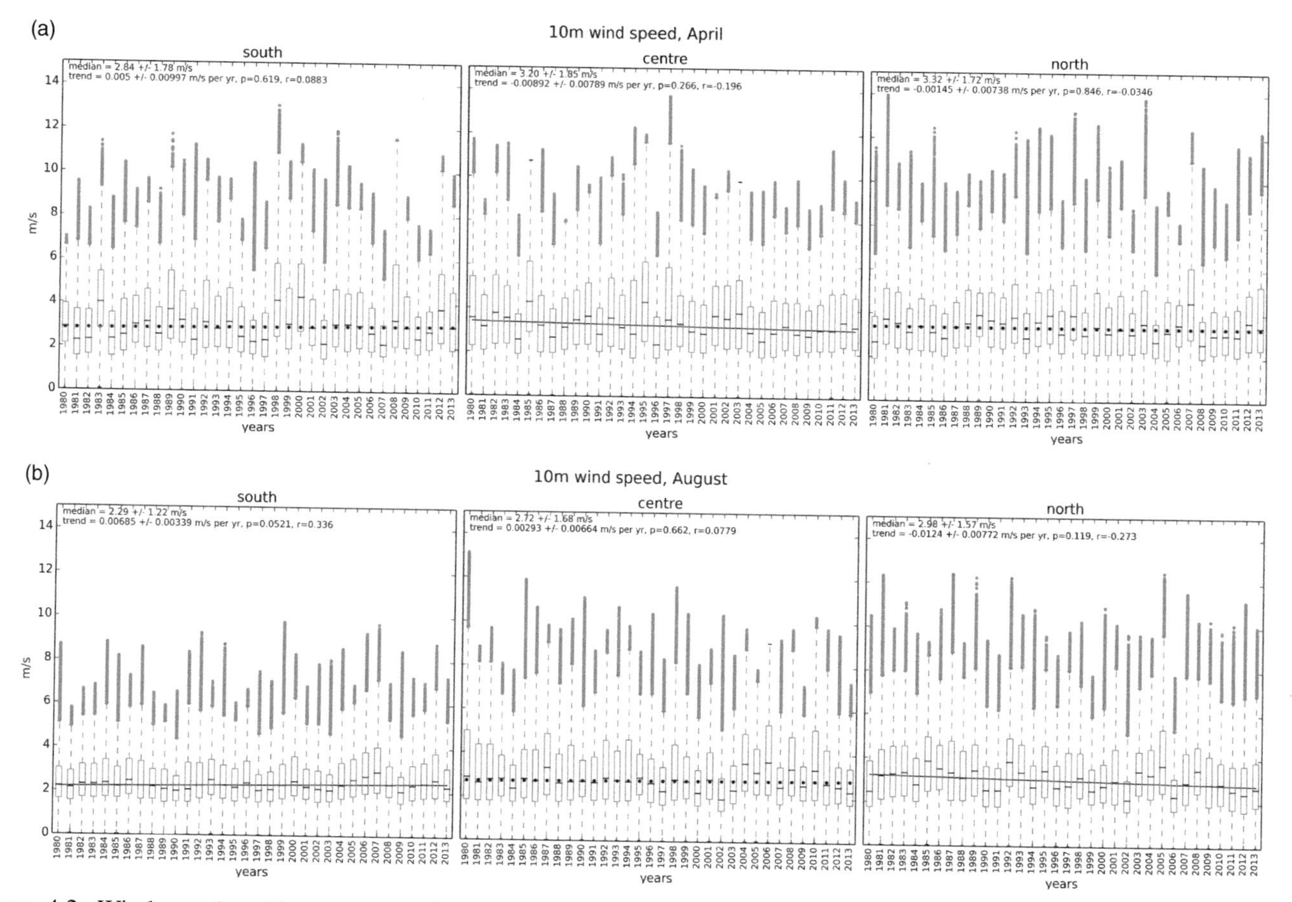

Figure 4.2. Wind speed at 10 m in (a) April and (b) August for the years 1980–2013 for three European regions: south (6°W, 38°N – 3°W, 41°N), central (10°E, 49°N – 13°E, 52°N), north (22°E, 61°N – 25°E, 64°N). Median and its trend are marked by red; data quartiles and outliers are blue. (Unit: m s^{-1}.) (A black and white version of this figure will appear in some formats. For the colour version, please refer to the plate section.)

The wind reduction may be opposed by stronger and more frequent storms, which, according to IPCC, will occur in middle latitudes (IPCC, 2013). However, it will probably still be too rare a phenomenon to significantly change the pollen dispersion pattern. Also, storms are usually accompanied by precipitation events, which will further reduce the promoting effect of strong winds on pollen transport: the atmosphere will be cleaned out by rain.

The implications of the projected gradual wind reduction, if it realises, are quite straightforward:

1. The transport velocity will become somewhat slower and pollen grains will travel shorter distances before settling down.
2. Less mechanically induced turbulence will be generated near the ground; that is, pollen mobilisation and resuspension will also decrease.
3. All in all, pollen concentrations will reduce far from the sources (over few tens of kilometres) but increase in the regions with major plant populations.

4.2.3 Transport Direction

Potential rearrangements in the atmospheric flows may result in alternative pathways of pollen dispersion at continental scale. The problem, however, is that to date there is no established dispersion patterns even for major pollen types in Europe. Numerous studies reported a large variety of episodes, each being practically unique for the region and for the year and not repeating itself in other regions and years (Damialis *et al.*, 2005; Hernandez-Ceballos *et al.*, 2014; Šikoparija *et al.*, 2009; Skjøth *et al.*, 2007, 2009; Smith *et al.*, 2005, 2008; Stach *et al.*, 2007; Veriankaitė *et al.*, 2010; see also chapter 5 of Sofiev and Bergmann, 2013, and references therein). There are just a few facts that seem to be established: (i) on average, the main transport direction is from west to east and, to a less extent, from south to north; (ii) across and counter to this main stream, intensive exchange of pollen is driven by large-scale cyclonic vortices; (iii) for short-time flowering species, such as birch and olive, the weather situation during the flowering days decides the annual pattern for the specific year. The later observation is the main reason for high variability of annual pollen dispersion patterns.

Generic consideration of the prevailing wind directions and transport patterns shows evident differences between the considered regions (Figure 4.3), but trends are quite uncertain. Formal analysis for April (Figure 4.3a) shows very wide variability around the prevailing west-to-east direction (zeroth degree). No major changes are noticeable (the apparent trend in the central region is not statistically significant; $p \sim 0.15$). Qualitative consideration of the scatter plot (Figure 4.3b, where the latitudinal wind component v is plotted against its longitudinal component u) shows some small increase of eastwards transport in the south and

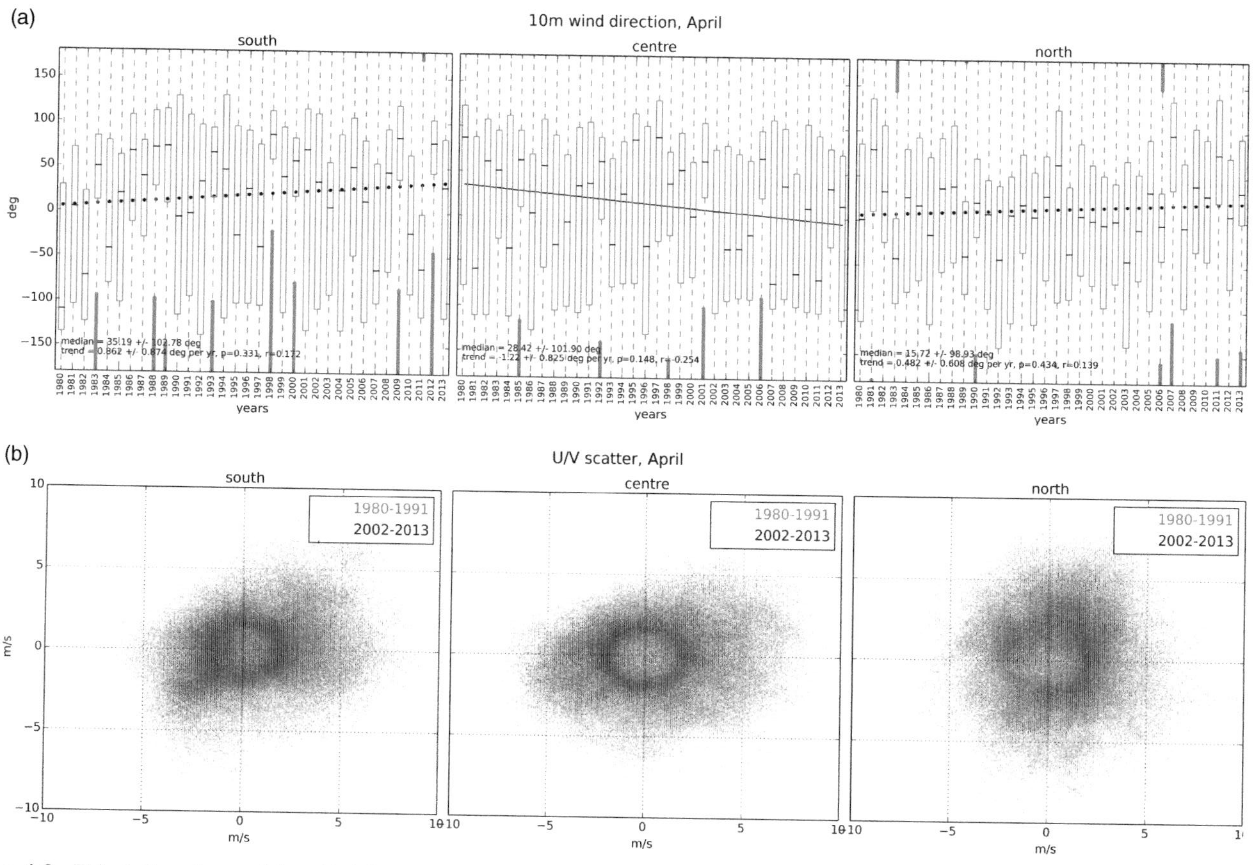

Figure 4.3. Wind direction at 10 m (ϕ_{10}) in April for the years 1980–2013 (a), and u_{10}–v_{10} scatter plots for wind at 10 m (b). Regions and notations are the same as in Figure 4.2. (Unit: degrees for (a) and m s^{-1} for (b).) (A black and white version of this figure will appear in some formats. For the colour version, please refer to the plate section.)

northeastwards in the central and northern sub-domains, but the main transport patterns in 1980s and 2010s are still similar.

4.2.4 Turbulence Intensity

The characteristics of mixing are usually not available from meteorological and climate models. They have to be diagnosed from surface sensible and latent heat fluxes (also not always available from climate models) or from the profiles of temperature and wind speed. For the current analysis, we used the approach of Sofiev *et al.* (2010), which is based on extended Monin–Obukhov similarity theory and provides the eddy diffusivity coefficient at 1 m height.

A significant drawback of this parameter is that some climate models tend to report near-neutral stratification in the lower part of the troposphere. Therefore, the projections of eddy diffusivity for the future climate have to be taken with great care. For a period of ~10 years, some information can be extrapolated from the ERA-Interim fields, but the uncertainty will be large.

From Figure 4.4, it is clear that this parameter is among the most stable ones: inter-annual variability is the smallest of all variables considered. Trends are insignificant in spring but small positive in the south and tentatively negative in the north – in summer (though barely significant; p ~0.14). The upward trend in the south is not surprising since turbulence is stronger on hot clear sky days. Therefore, this trend will continue with further heating of the region in the future. As a result, summer-time flowering species will get more favourable conditions for pollen to be picked up from inflorescences and mixed within the whole boundary layer.

4.2.5 Relative Humidity

The boundary layer relative humidity is important for pollen release because the inflorescences of many species close at high humidity. During the course of transport, pollen exchanges water with the surrounding air and its aerodynamic size grows in humid air and shrinks in dry conditions. As a result, the pollen settling velocity can substantially change, which in turn would affect its dry deposition intensity and lifetime in the atmosphere.

There is a peculiarity of the relative humidity parameter: it reflects the combination of actual water content in the air (specific humidity expressed as mass of water per unit mass of air) and air temperature. In warmer air, the same relative humidity means higher specific humidity, i.e., additional water vapour present in the atmosphere. Practically all climate studies agree that specific humidity will rise at a rate of 3–6% per degree of global warming, as evaporation from vegetation and water surfaces will increase with the rising temperatures. This is one of the positive

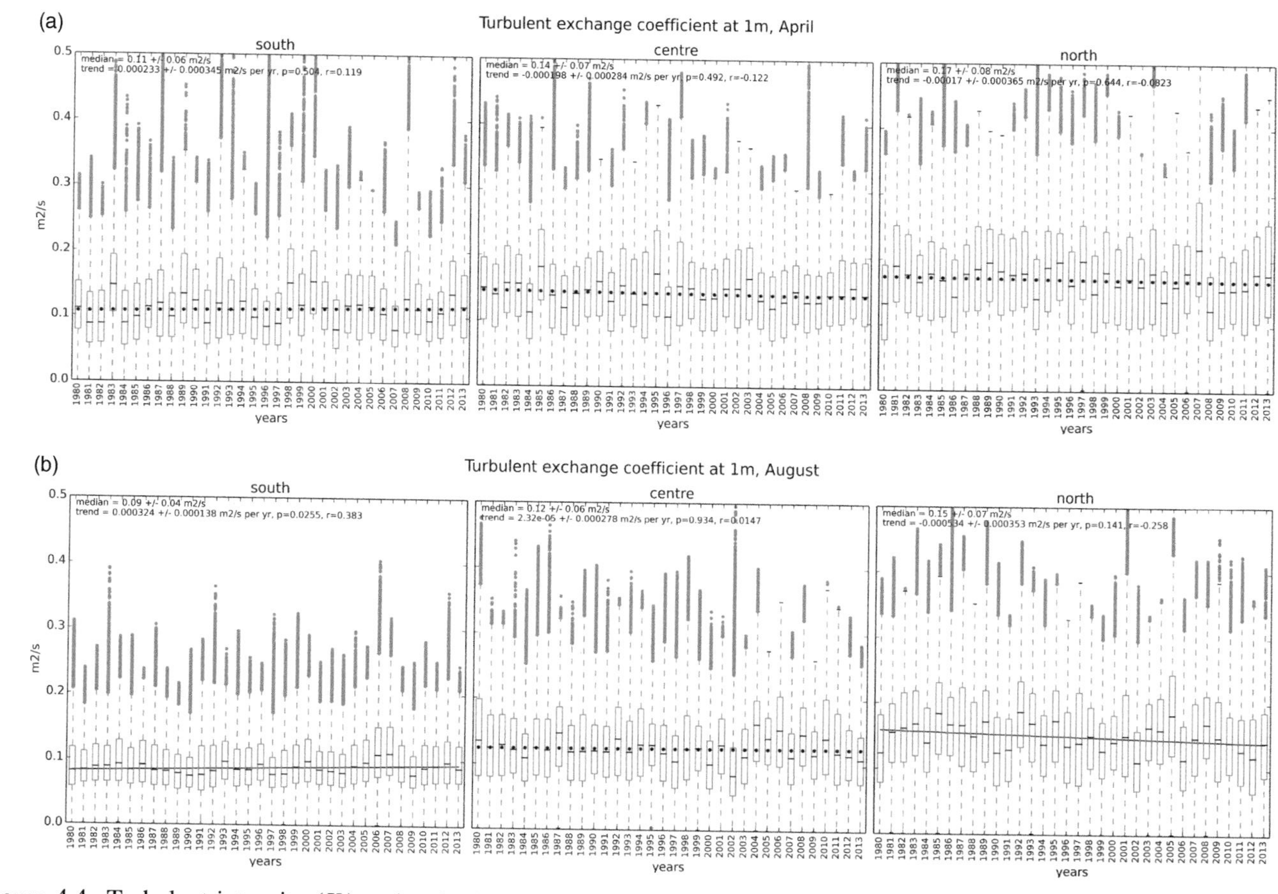

Figure 4.4. Turbulent intensity (K_z) at 1 m in (a) April and (b) August for the years 1980–2013. Regions and notations are the same as in Figure 4.2. (Unit: $m^2 s^{-1}$.) (A black and white version of this figure will appear in some formats. For the colour version, please refer to the plate section.)

feedback mechanisms of climate change: water vapour itself is a greenhouse gas and its abundance would increase warming. The rise of specific humidity has a substantial latitudinal trend: in Southern Europe, the increase will be 2–5% in the near term (mean for 2016–2035 compared to the mean for 1986–2005), whereas up to 10% increase can be expected in the north (Kirtman *et al.*, 2013). Relative humidity, on the other hand, is projected to slightly decrease, especially in the south due to limited amounts of evaporable water – but the changes are just 1–2% and variability and disagreement between the models are larger than the projected changes.

From Figure 4.5, one can see that European spring-time relative humidity has no visible trend during the last 33 years in the south but decreased in the central and, with lower significance, northern regions. In August, the situation is the opposite: already low humidity level in the south has strong negative trend, whereas other regions appeared completely stable. This observation is a good illustration of inhomogeneity of climate change – an expectation that the south of the continent will dry is not observed in spring: in fact, the only season in the south that is getting substantially drier is summer. In winter, relative humidity in the south is even increasing – by ~0.1% per year.

Decreasing mean relative humidity is likely to cause longer pollen transport because the drier grains will have lower sedimentation velocity. It will also be easier for the plants to release pollen immediately as it matures. The effects will be most pronounced in the south of Europe during summer and, if observed trends continue, in central and northern regions of Europe during spring time.

4.2.6 Fraction of Precipitation Periods

Comparison of the current and projected future trends of total precipitation amount per year and frequency of precipitation events reveals a complicated picture with very large variability between the years and uncertain trends. According to IPCC (Hartmann *et al.*, 2013), the 1950–2010 trends in annual precipitation amount were positive over central and northern Europe (roughly north of 45°N, up to 10 mm yr^{-1} per decade) and negative in the south (south of 45°N, down to –25 to –50 mm yr^{-1} per decade). There was also a general tendency to have longer and stronger droughts; that is, the rain events tend to group into shorter but stronger showers with longer dry spells in-between.

The ERA-Interim reanalysis for April (Figure 4.6a) also shows a complicated picture. In the southern region, two distinct sub-periods can be noticed: 1980–2005 with a gradually decreasing fraction of rain periods, and the last 8 years when this fraction grows. The second period roughly coincides with the 'pause' in the rise of global mean temperature that started in early 2000s (however, all other main

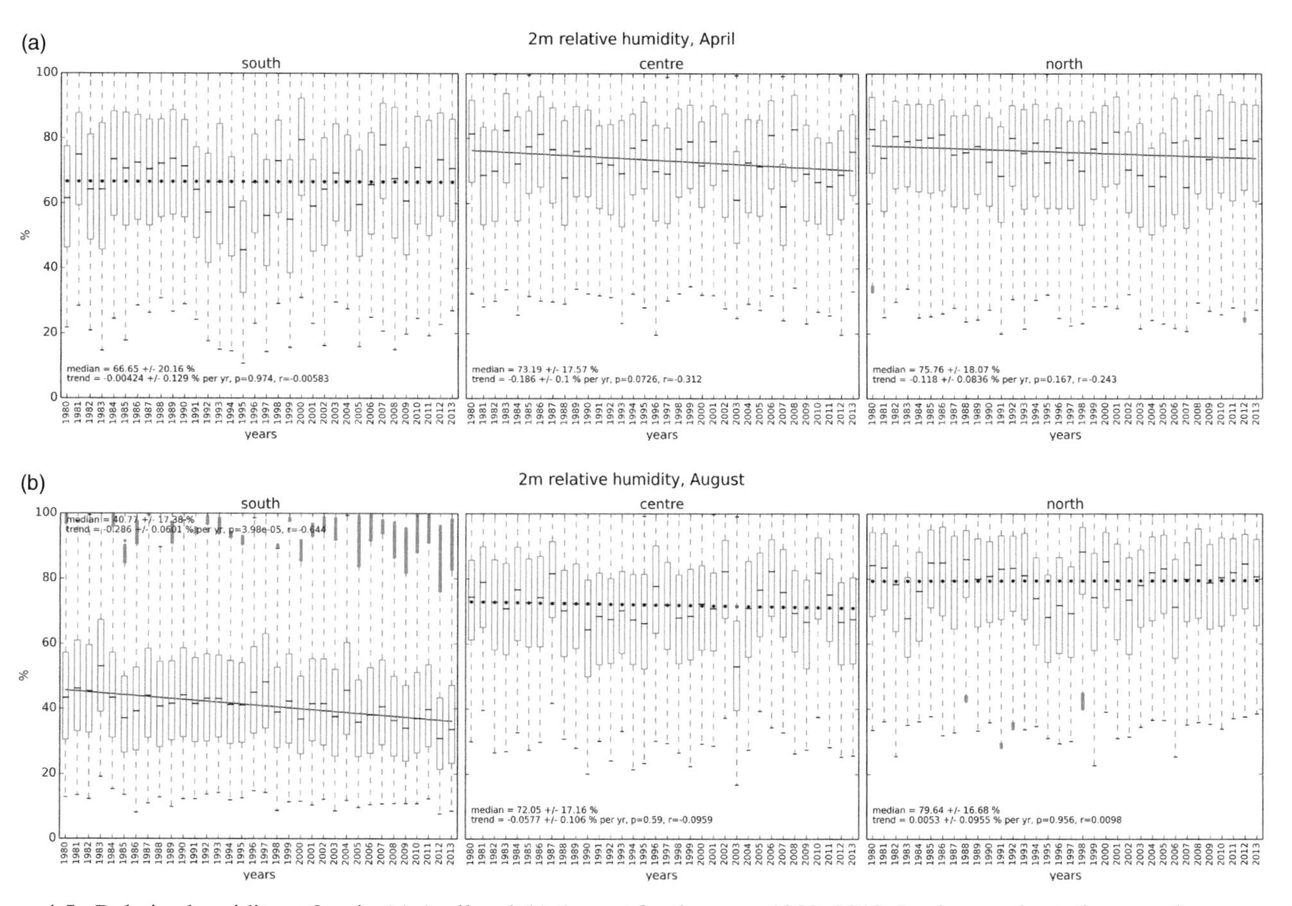

Figure 4.5. Relative humidity at 2 m in (a) April and (b) August for the years 1980–2013. Regions and notations are the same as in Figure 4.2. (Unit: percentage.) (A black and white version of this figure will appear in some formats. For the colour version, please refer to the plate section.)

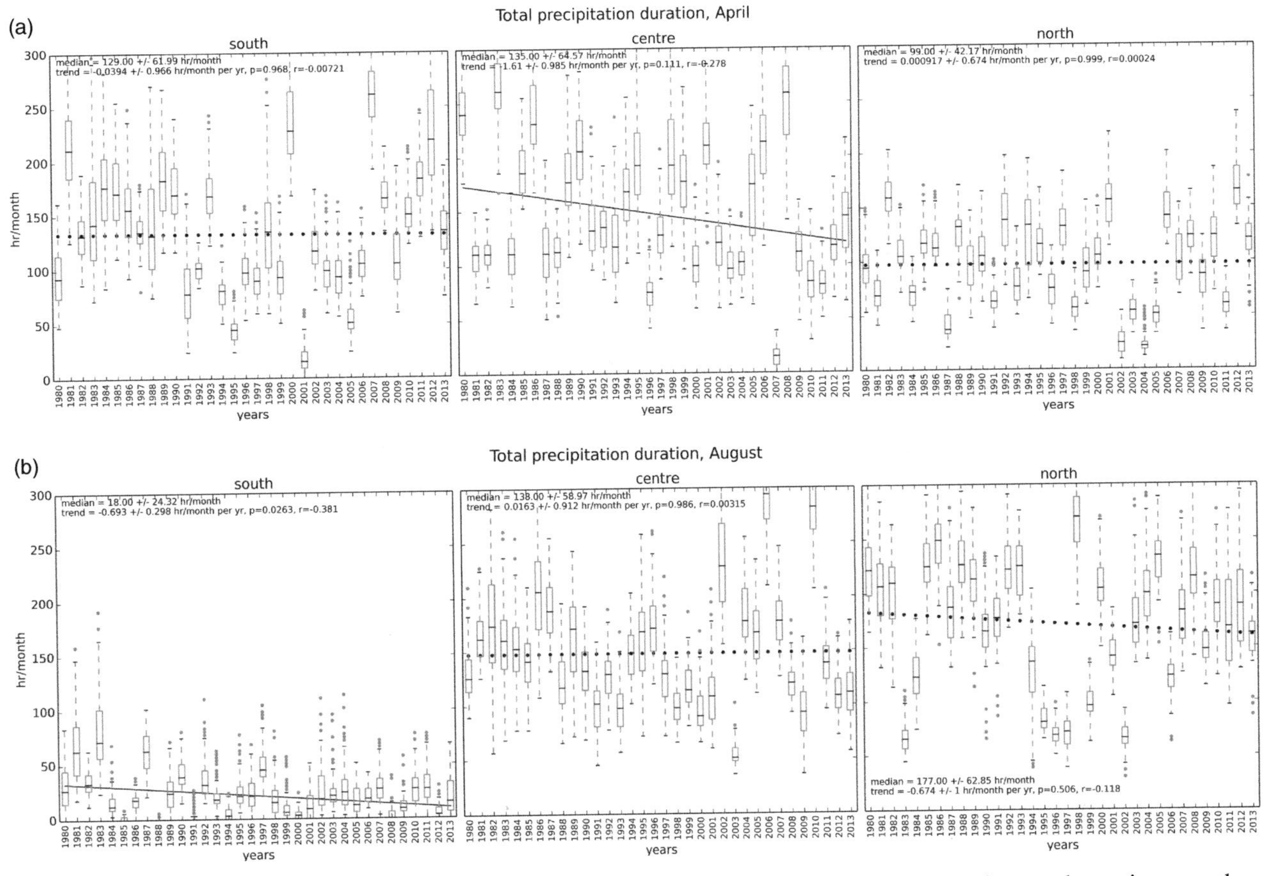

Figure 4.6. Total monthly rain duration in (a) April and (b) August for the years 1980–2013. Regions and notations are the same as in Figure 4.2. (Unit: hour.) (A black and white version of this figure will appear in some formats. For the colour version, please refer to the plate section.)

climate characteristics continued the trends – see Stocker *et al.*, 2013), but it is too short a period to make any extrapolations. Should the complete period of 33 years be considered, the trend is not significant. In central Europe, despite a very large scatter, a negative trend is visible, largely reflecting long dry periods in April since ~2000 (with just three exceptions – Figure 4.6a). There is no trend in the north.

In August (Figure 4.6b), the situation looks very stable though different among the regions. There is essentially no rain in the south; its median duration is stable at 20–30 rainy hours per month. The apparent trend is due to the disappearance of rainy months, which were sporadically seen in the 1980s but never since then. The central region does not show any trend either but exhibits both anomalously dry and wet months more frequently after 2000. The opposite is the case in the north, where the last 10 years were stably wet with just one exception in 2006. The series of dry summers of the mid-1990s looks as an isolated case.

Longer dry periods with stronger but shorter rains in-between will promote pollen transport, thus working in the same direction as the reduction in relative humidity.

4.3 Climate Change Impact on Air Pollutants: What Can We Learn?

There are several model-based studies considering the impact of climate change on various pollutants, such as ozone, nitrogen oxides, particulate matter, etc. Few of them considered non-reactive aerosols – and they practically never considered the primary biogenic particles. In all studies considering the chemically reactive aerosol, the climate change impact via chemical transformations and changes of formation or emission rates by far overwhelmed the direct effect of weather on particles. IPCC came up with the following conclusion: 'The range in projections of air quality (O_3 and $PM_{2.5}$ in near-surface air) is driven primarily by emissions (including CH_4), rather than by physical climate change (medium confidence). The response of air quality to climate-driven changes is more uncertain than the response to emission-driven changes (high confidence)' (Kirtman *et al.*, 2013). However, one can point out that coarse aerosols, being potentially more susceptible to weather phenomena than fine particles, were left out of the consideration.

Other effects pointed out by the authors include the shift of thermodynamic equilibrium of some aerosol species (Liao *et al.*, 2006) and potential increase of some sources of aerosols, such as biomass burning (Jacob and Winner, 2009). However, some tendencies important also for non-reactive aerosols were also highlighted (Horton *et al.*, 2014; Isaksen *et al.*, 2009).

Probably the most important effect will be due to changes in the hydrological cycle and wet removal processes in some regions, generally increasing scavenging of constantly present species in the atmosphere (Liao *et al.*, 2006; Racherla and

Adams, 2006). Second, higher levels of oxidants will tend to increase secondary aerosol formation rates (Stevenson *et al.*, 2006; Tsigaridis and Kanakidou, 2007; Unger *et al.*, 2006) – and also cause more damage to the pollen grains during the transport. Third, potential increase of stagnation events (Horton *et al.*, 2014) will limit the pollen transport distance and increase its concentrations in the vicinity of the sources.

Bulk estimates suggest overall reduction of aerosol load. Thus, global aerosol burden is reduced by 20% during 100 years (2000–2100) in the study of Liao *et al.* (2006), and by 2–18% during 60 years (1990–2050), as concluded by Racherla and Adams (2006), mainly due to enhanced deposition.

A common feature of the bulk of such studies is that the changes are considered at annual level, rarely for winter/summer half-years, while for most plant species the period of flowering is short (weeks, rarely 1–2 months), and it is the dominating weather during this time that is important. In addition, many wind-pollinated plants are well prepared to wait through wet days, no matter how strong rains will be, and release pollen only in dry hot windy periods, which will become longer. The question therefore is again on frequency and duration of such good days rather than on general change of the mean annual values of some weather characteristics.

4.4 Examples of Changes of the Pollen Transport Patterns

This section presents the results of a multi-decade pollen dispersion study for past climate. It is mainly based on the material of a forthcoming paper (M. Sofiev *et al.*, in preparation).

The study used the System for Integrated modeLing of Atmospheric coMposition (SILAM; http://silam.fmi.fi) and its pollen source terms (Prank *et al.*, 2013; Siljamo *et al.*, 2013; Sofiev *et al.*, 2006, 2013). Below we present the study results for two taxa: birch, as an example of a spring-time flowering tree with significant long-range transport of pollen, and grasses, as an example of plants with a long flowering season and short distance of pollen transport. Pollen emission of each taxon was described with its own phenological model: double-thermal sum for birch and a simple fixed-dates approach for grass. The release of pollen to the atmosphere was dependent on meteorological conditions, such as temperature, relative humidity, precipitation, and wind speed. Other processes included in SILAM cover the whole atmospheric lifecycle of pollen after its release into the air: transport with air masses, turbulent mixing, and removal by dry and wet deposition. Pollen transport and removal account for the processes governing the dispersion of chemically non-reactive aerosols, depending on the basic properties of the pollen grains, such as their aerodynamic diameter and density (Kouznetsov and Sofiev, 2012).

The maps of total emitted annual pollen amounts were kept constant; that is, no vegetation responses to the meteorological situation before the flowering season or adaptation to the changing climate was considered.

SILAM was run through the past 33 years using the ERA-Interim meteorological archive analysed in previous sections, revealing the trends connected with the ongoing climate change. Obtained trends have been compared with the analysis of the previous sections.

Technical setup of the model runs was as follows: the simulations covered the whole of Europe with spatial resolution of 0.3 × 0.2 degrees in longitude–latitude directions, respectively, and eleven layers stacked up to about 6.5 km. The internal model time step was 15 minutes with hourly output of surface concentrations and basic meteorological parameters.

4.4.1 Birch: Spring-Time Wind-Pollinated Tree

The outcome of the simulations for birch is shown in Figure 4.7. Comparison of these maps with each other reveals certain correlation between the load and the trend: pollen tends to stay closer to the sources in later years. The mean increase during 1980–2012 was more than 100 pollen day m^{-3} yr^{-1}, i.e., 0.1% per year – over large territories in Russia and Baltic States, southern Sweden, and northwestern Norway (though it is a complex terrain area and current computations have much too crude resolution for it).

Over Finland and northern Russia, trends are mainly negative, except for western Finland, and again very substantial: ~–0.2% per year on average over 30 years.

This result supports the suggestion of the previous section that ongoing climate change will reduce the transport distance of the species, owing to combined effect of wind speed reduction and changes in precipitation regime. The controversial trends in central Europe and in the north are then explained by difference in the trends of precipitation pattern: reduction of rainy time in April in Germany (Figure 4.6a) and increase of the precipitation duration in May in Finland (the main birch flowering month there, not shown).

4.4.2 Grass: Large Pollens with Short Transport Distance

Grass pollen is usually considered as a local-scale aerosol, owing to its size (1.5–2 times that of birch pollen) and low emission height, which makes it difficult to raise the grains up to the dispersion layers. Therefore, one can expect that wind changes would be of minor importance for it, whereas the local-scale features, such as relative humidity and a fraction of rainy days, would control the concentrations.

The SILAM simulations (Figure 4.8) showed a pretty homogeneous increase in all pollinating regions: there were essentially no areas with significant negative

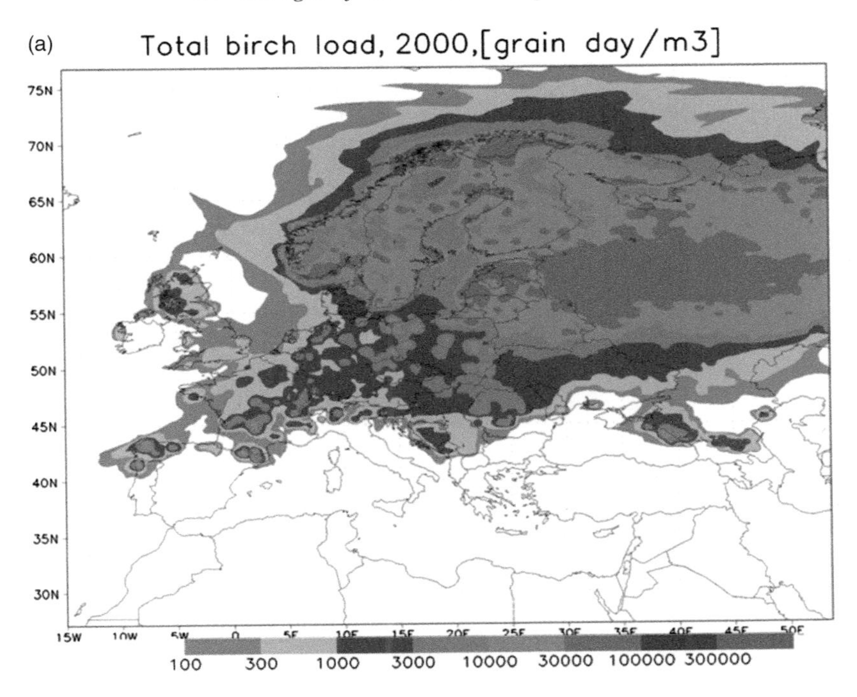

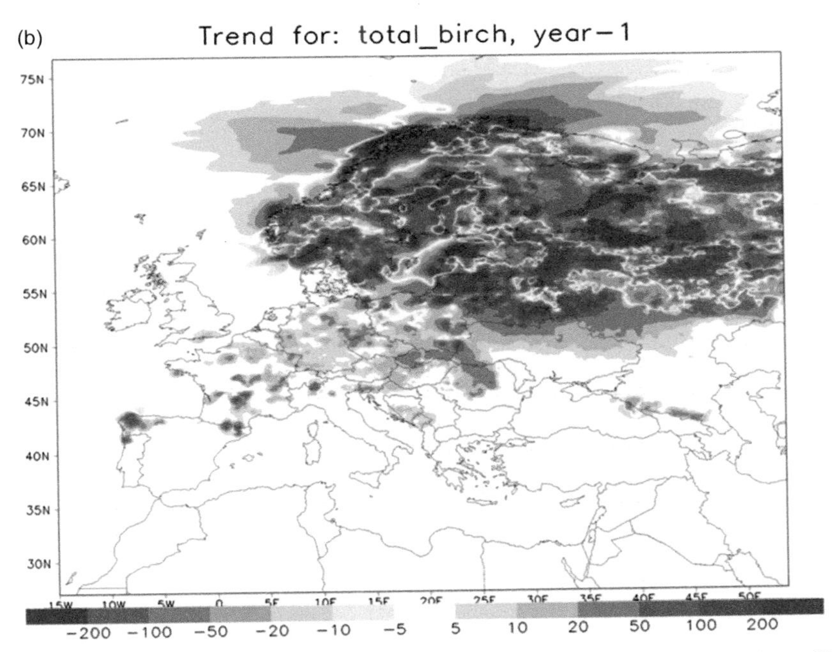

Figure 4.7. Birch total seasonal pollen count in Europe for 2000 (a; pollen day m^{-3}) and its 1980–2012 trend (b; pollen day m^{-3} yr^{-1}). (A black and white version of this figure will appear in some formats. For the colour version, please refer to the plate section.)

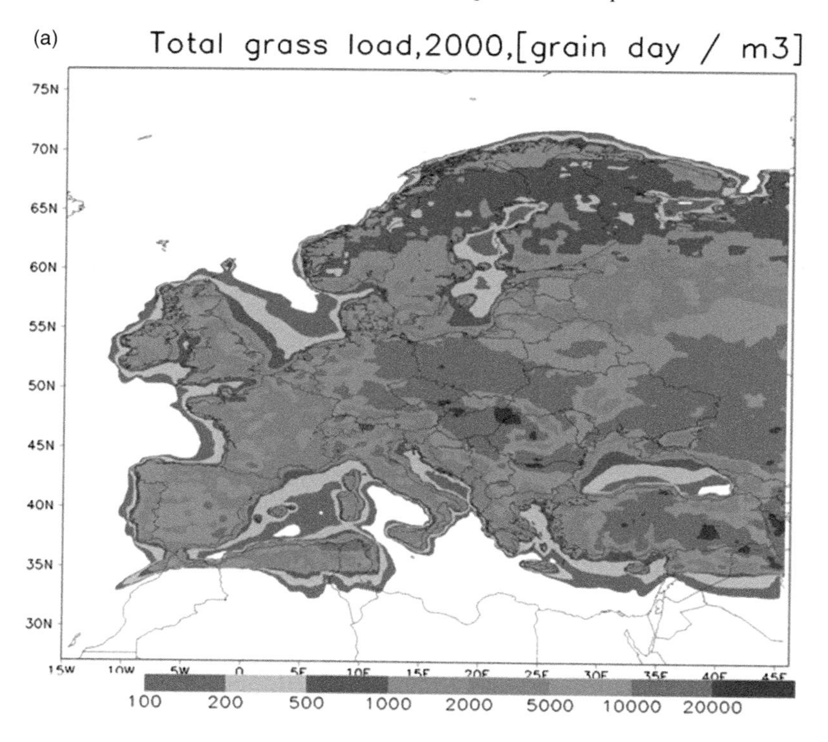

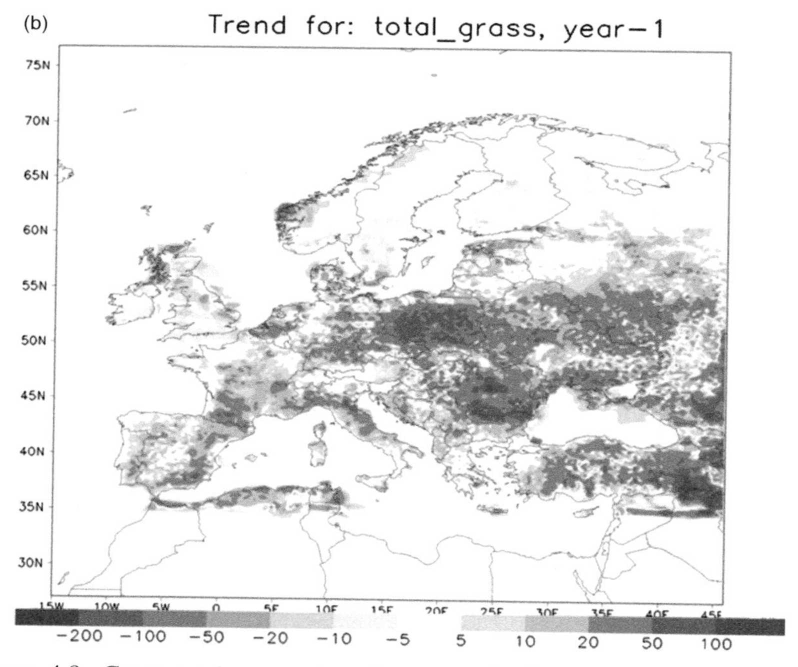

Figure 4.8. Grass total seasonal pollen count in Europe for 2000 (a; pollen day m^{-3}) and its 1980–2012 trend (b; pollen day m^{-3} yr^{-1}). (A black and white version of this figure will appear in some formats. For the colour version, please refer to the plate section.)

trends. The growth again mounted up to ~0.1% per year. This growth should probably be attributed to reduction of the fraction of the rainy days in May–July.

4.5 Uncertainties in the Current Knowledge

Estimating the climate change impact on pollen dispersion and deposition is a somewhat ambiguous exercise because this impact is not observable: meteorology affects the plant habitat and pollen production much more than its transport in the air. Indeed, the main parameter considered for climate change – temperature – is practically irrelevant for transport but largely decides on flowering characteristics. Precipitation events scavenge out the grains from the air but also prevent the release of fresh portions from inflorescences, while droughts can kill the plants and end the flowering season. Wind facilitates the release by picking up pollen but also cleans up the atmosphere in the source region pushing the grains away and reducing the near-plant concentrations. Aerobiological observations register the net effect of all these processes – the actual pollen concentrations – but they cannot distinguish between the specific phenomena responsible for the changes.

Model-based assessments face different types of problems: sensitivity runs (when only one parameter or input dataset is changed) can reveal the impact of each factor, but it is hard to demonstrate that the obtained results have connection to reality. Indeed, the current stage of pollen modelling is characterised by a large number of various empirical parameterisations, which may be useful to describe the current situation but by no means proven to work in different climatic conditions. This is particularly severe for the source term descriptions, which are all entirely empirical. Pollen behaviour in the atmosphere is established better: basic physical laws are valid regardless of the atmospheric conditions, and their current representation in dispersion models is sufficient for pollen calculations (Sofiev *et al.*, 2006). Therefore, transport with wind, mixing, and deposition can be analysed with high confidence. Much less is known about the water budget of the airborne pollen grain and its water exchange with the surrounding air. To the best of our knowledge, they are ignored in the existing models, thus making it impossible to analyse the trends due to changing hydrological cycle and air humidity.

Direct pollen dispersion modelling also faces the third type of obstacle: inhomogeneity of the input data. Indeed, the ERA-Interim reanalysis, albeit being arguably the best dataset in the class, contains a major uncertainty connected with a strongly varying amount of observations available for earlier years. As stated by Dee *et al.* (2011), the amount of assimilated information grew from about 10^6 per day in 1989 to nearly 10^7 per day in 2010. The bulk of the increase came from satellite observations, which were also an overwhelming majority of the input data. In application to the current work, the biggest concern comes from inhomogeneous flow of data on precipitable water in the air column, which in the early 1990s constituted barely 10^3 entries per day, exceeded 2×10^4 in the mid-2000s, fell below 7×10^3 in 2009, and was not available in 2010. Nevertheless, the impact of this inhomogeneity is not visible in the trend analysis, primarily because in Europe the density of conventional observations is anyway very high and steady. Therefore, for the considered regions one can still expect sufficiently homogeneous quality of the information and, consequently, reliability of the reported trends.

4.6 Summary

Keeping in mind the limitations of the current knowledge, we focussed on two types of analysis: (i) trends of meteorological parameters that are expected to affect the pollen transport, and (ii) direct pollen dispersion modelling with fixed source term formulations and parameters.

Analysis of meteorological components for the future climate, largely based on IPCC information and ERA-Interim reanalysis, suggested the following trends:

- Reduction of mean wind speed, with simultaneous empowering of storms; these trends are weak in Europe but in a few cases visible from ERA-Interim data.
- Increase of specific humidity with uncertain and diverse trends of relative humidity; ERA-Interim dataset suggested moderate reduction of spring-time relative humidity in central and northern Europe but significant reduction in summer in the south.
- Increase of duration and severity of dry spells, with simultaneous increase of power of individual rains, so that the total precipitation amount grows; this tendency is visible from ERA-Interim in the south in summer and, with high variability, in central Europe in spring.

The overall effect of these trends should be a reduction of the pollen transport distance causing reduction of concentrations away from the sources. Changes in the near fields are more uncertain due to competition and partial mutual offsetting of these factors. This conclusion is in general agreement with studies published for 'conventional' pollutants.

Pollen dispersion modelling over last 33 years generally confirmed the trends and suggested, in most cases, that the near-source concentrations grow by ~0.1% per year. A reduction of similar magnitude was suggested for the regions away from the main sources, with a chance to become even more pronounced when the effect of reducing relative humidity is included in the model.

Acknowledgements

Pollen source terms of SILAM model have been developed in cooperation with European Aeroallergen Network within the scope of the projects POLLEN and

APTA of Academy of Finland, and FP7 MACC, HIALINE, and ENV.B2/ETU/2010/0037 ('Assessing and controlling the spread and the effects of common ragweed in Europe').

References

Damialis, A., Gioulekas, D., Lazopoulou, C., Balafoutis, C., Vokou, D. (2005). Transport of airborne pollen into the city of Thessaloniki: the effects of wind direction, speed and persistence. *International Journal of Biometeorology*, 49(3), 139–145.

Dee, D. P., Uppala, S. M., Simmons, A. J., *et al.* (2011). The ERA-Interim reanalysis: configuration and performance of the data assimilation system. *Quarterly Journal of the Royal Meteorological Society*, 137(656), 553–597.

Di-Giovanni, F., Kevan, P. G. (1991). Factors affecting pollen dynamics and its importance to pollen contamination: a review. *Canadian Journal of Forest Research*, 21(8), 1155–1170.

Hartmann, D. L., Klein Tank, A. M. G., Rusticucci, M., *et al.* (2013). Observations: atmosphere and surface. In: Stocker, T. F., Qin, D., Plattner, G.-K., *et al.*, eds. *Climate Change 2013: The Physical Science Basis. Contribution of Working Group I to the Fifth Assessment Report of the Intergovernmental Panel on Climate Change*. Cambridge, UK and New York, NY, USA: Cambridge University Press, pp. 159–254.

Helbig, N., Vogel, B., Vogel, H., Fiedler, F. (2004). Numerical modelling of pollen dispersion on the regional scale. *Aerobiologia*, 20(1), 3–19.

Hernandez-Ceballos, M. A., Soares, J., García-Mozo, H., *et al.* (2014). Analysis of atmospheric dispersion of olive pollen in southern Spain using SILAM and HYSPLIT models. *Aerobiologia*, 30(3), 239–255.

Horton, D. E., Skinner, C. B., Singh, D., Diffenbaugh, N. S. (2014). Occurrence and persistence of future atmospheric stagnation events. *Nature Climate Change*, 4, 698–703.

IPCC (2013). *Climate Change 2013: The Physical Science Basis. Contribution of Working Group I to the Fifth Assessment Report of the Intergovernmental Panel on Climate Change* (Stocker, T. F., Qin, D., Plattner, G.-K., *et al.*, eds.). Cambridge, UK: Cambridge University Press.

Isaksen, I. S. A., Granier, C., Myhre, G., *et al.* (2009). Atmospheric composition change: climate–chemistry interactions. *Atmospheric Environment*, 43(33), 5138–5192.

Jacob, D. J., Winner, D. A. (2009). Effect of climate change on air quality. *Atmospheric Environment*, 43(1), 51–63.

Kirtman, B., Power, S. B., Adedoyin, J. A., *et al.* (2013). Near-term climate change: projections and predictability. In: Stocker, T. F., Qin, D., Plattner, G.-K., *et al.*, eds. *Climate Change 2013: The Physical Science Basis. Contribution of Working Group I to the Fifth Assessment Report of the Intergovernmental Panel on Climate Change*. Cambridge, UK and New York, NY, USA: Cambridge University Press, pp. 953–1028.

Kouznetsov, R., Sofiev, M. (2012). A methodology for evaluation of vertical dispersion and dry deposition of atmospheric aerosols. *Journal of Geophysical Research: Atmospheres*, 117(D01), D01202.

Liao, H., Chen, W.-T., Seinfeld, J. H. (2006). Role of climate change in global predictions of future tropospheric ozone and aerosols. *Journal of Geophysical Research: Atmospheres*, 111(D12), D12304.

Linskens, H. F., Cresti, M. (2000). Pollen-allergy as an ecological phenomenon: a review. *Plant Biosystems*, 134(3), 341–352.

McVicar, T. R., Roderick, M. L., Donohue, R. J., *et al.* (2012). Global review and synthesis of trends in observed terrestrial near-surface wind speeds: implications for evaporation. *Journal of Hydrology*, 416–417, 182–205.

Prank, M., Chapman, D. S., Bullock, J. M., *et al.* (2013). An operational model for forecasting ragweed pollen release and dispersion in Europe. *Agricultural and Forest Meteorology*, 182–183, 43–53.

Racherla, P. N., Adams, P. J. (2006). Sensitivity of global tropospheric ozone and fine particulate matter concentrations to climate change. *Journal of Geophysical Research: Atmospheres*, 111(D24), D24103.

Šikoparija, B., Smith, M., Skjøth, C. A., *et al.* (2009). The Pannonian plain as a source of *Ambrosia* pollen in the Balkans. *International Journal of Biometeorology*, 53(3), 263–272.

Siljamo, P., Sofiev, M., Filatova, E., *et al.* (2013). A numerical model of birch pollen emission and dispersion in the atmosphere. Model evaluation and sensitivity analysis. *International Journal of Biometeorology*, 57(1), 125–136.

Skjøth, C. A., Smith, M., Brandt, J., Emberlin, J. (2009). Are the birch trees in Southern England a source of *Betula* pollen for North London? *International Journal of Biometeorology*, 53(1), 75–86.

Skjøth, C. A., Sommer, J., Stach, A., Smith, M., Brandt, J. (2007). The long-range transport of birch (*Betula*) pollen from Poland and Germany causes significant pre-season concentrations in Denmark. *Clinical and Experimental Allergy*, 37(8), 1204–1212.

Smith, M., Emberlin, J., Kress, A. (2005). Examining high magnitude grass pollen episodes at Worcester, United Kingdom, using back-trajectory analysis. *Aerobiologia*, 21(2), 85–94.

Smith, M., Skjøth, C. A., Myszkowska, D., *et al.* (2008). Long-range transport of *Ambrosia* pollen to Poland. *Agricultural and Forest Meteorology*, 148(10), 1402–1411.

Sofiev, M., Bergmann, K.-C., eds. (2013). *Allergenic Pollen. A Review of the Production, Release, Distribution and Health Impacts.* Dordrecht: Springer.

Sofiev, M., Genikhovich, E., Keronen, P., Vesala, T. (2010). Diagnosing the surface layer parameters for dispersion models within the meteorological-to-dispersion modeling interface. *Journal of Applied Meteorology and Climatology*, 49(2), 221–233.

Sofiev, M., Siljamo, P., Ranta, H., *et al.* (2013). A numerical model of birch pollen emission and dispersion in the atmosphere. Description of the emission module. *International Journal of Biometeorology*, 57(1), 45–58.

Sofiev, M., Siljamo, P., Ranta, H., Rantio-Lehtimäki, A. (2006). Towards numerical forecasting of long-range air transport of birch pollen: theoretical considerations and a feasibility study. *International Journal of Biometeorology*, 50(6), 392–402.

Stach, A., Smith, M., Skjøth, C. A., Brandt, J. (2007). Examining *Ambrosia* pollen episodes at Poznań (Poland) using back-trajectory analysis. *International Journal of Biometeorology*, 51(4), 275–286.

Stevenson, D. S., Dentener, F. J., Schultz, M. G., *et al.* (2006). Multimodel ensemble simulations of present-day and near-future tropospheric ozone. *Journal of Geophysical Research: Atmospheres*, 111(D8), D08301.

Stocker, T. F., Qin, D., Plattner, G.-K., *et al.* (2013). Technical summary. In: Stocker, T. F., Qin, D., Plattner, G.-K., *et al.*, eds. *Climate Change 2013: The Physical Science Basis. Contribution of Working Group I to the Fifth Assessment Report of the Intergovernmental Panel on Climate Change*. Cambridge, UK and New York, NY, USA: Cambridge University Press, pp. 33–115.

Tsigaridis, K., Kanakidou, M. (2007). Secondary organic aerosol importance in the future atmosphere. *Atmospheric Environment*, 41(22), 4682–4692.

Unger, N., Shindell, D. T., Koch, D. M., Streets, D. G. (2006). Cross influences of ozone and sulfate precursor emissions changes on air quality and climate. *Proceedings of the National Academy of Sciences of the United States of America*, 103(12), 4377–4380.

Veriankaitė, L., Siljamo, P., Sofiev, M., Šaulienė, I., Kukkonen, J. (2010). Modelling analysis of source regions of long-range transported birch pollen that influences allergenic seasons in Lithuania. *Aerobiologia*, 26(1), 47–62.

5

Impacts of Climate Change on Allergenicity

JEROEN T. M. BUTERS[1,2]
[1]*Center of Allergy and Environment (ZAUM)*
Technical University Munich and Helmholtz Center Munich
[2]*Christine Kühne Center for Allergy Research and Education*

This chapter focusses on the impacts of climate change (e.g., increased air temperature and atmospheric carbon dioxide (CO_2) concentration) on allergenicity of pollen, mould spores, contact allergens, and plant food allergens. Temperature, humidity, CO_2, and thunderstorms are all changing due to climate change, and the effects on allergen release from sources are discussed.

5.1 Climate Change and Allergen Exposure

There is no doubt that climate change is taking place (IPCC, 2014). Many sources of allergens are of plant origin, especially pollen, which will react to changed growth conditions. Chemical allergens from inorganic sources like nickel in contact dermatitis, or allergens from warm-blooded animals like horses (Mitlehner, 2013) and cats (Kelly *et al.*, 2012), will show little change due to climate change. The effects of climate change on plant allergens are a consequence of changed growth control factors, not only temperature (global warming) but also light (changed albedo) or the rise in CO_2. CO_2 is driving climate change but is also a natural fertiliser, stimulating plant growth and differentiation and consequently allergenicity (Mohan *et al.*, 2006; Singer *et al.*, 2005). Evidence indicates that the effects of CO_2 are more influential than only the rise in temperature (Ziello *et al.*, 2012).

It is known that different climates result in differences in allergic disease (Asher *et al.*, 2006; Hesselmar *et al.*, 2001). Here we have to differentiate between 'becoming allergic', called sensitisation, and 'being already allergic' and having different symptoms due to changed exposure as a result of climate change. When being sensitised (becoming allergic) we have to differentiate between positive immune markers, almost always specific immunoglobulin E (IgE) antibodies but no symptoms, and having these antibodies but with symptoms of an allergic disease.

To have allergic symptoms, one needs exposure to an allergen, mostly a protein or glycoprotein (Traidl-Hoffmann *et al.*, 2009). Then an immune response is mounted. The hallmark of an allergic disease is specific IgE (sIgE) antibodies against one specific protein in the blood of the patient. Seldom do individuals have allergic reactions of the type I, commonly known as hay fever, without having sIgE antibodies.

In the literature on epidemiological studies, mostly the marker of allergic disease is reported, i.e., sIgE against one specific protein, e.g., the major birch pollen allergen Bet v 1. This has the advantage of not having to ask the patients whether they have symptoms or not, which is a parameter prone to show bias. Contrary to this, the laboratory marker of allergic disease (i.e., sIgE) is non-biased. However, about 50% of the positive laboratory values do not result in symptoms, i.e., are false-positive (Tschopp *et al.*, 1998). This is due to the polyclonal nature of immune globulins. Each immunoglobulin clone has a different affinity for, or epitope on, the antigen, e.g., Bet v 1 (Gieras *et al.*, 2011). Only when sufficiently specific high-affinity antibodies against an antigen are present in the patient's blood does mast cell degranulation occur (Willumsen *et al.*, 2012), resulting in the allergic symptoms. This hypothesis could explain why only about 50% of the individuals with positive sIgE have symptoms, although other hypotheses explaining this phenomenon like the induction of interleukin-10 are equally valid (Meiler *et al.*, 2008). Nevertheless, when reading the literature, one needs to differentiate between having a positive, objective blood marker IgE with a 50% predictive value and actually having an allergic disease. As a rule of thumb, 50% of the sIgE-positive values do result in allergic disease.

Interestingly, one study in Sweden (Hesselmar *et al.*, 2001) has found that between different climatic zones the number of allergic sensitisations (measured as sIgE) was different, i.e., implying more allergic disease in one climate zone than the other, but then the number of allergic diseased individuals between the different climatic zones were similar (Hesselmar *et al.*, 2001). The authors explained this by differences in dose. According to the authors, higher doses resulted in a natural desensitisation (sensitisation but no symptoms).

A worldwide study on allergic diseases (International Study of Asthma and Allergies in Childhood) also shows large differences in allergic sensitisations between different climates (Asher *et al.*, 2006). These results cannot be used to study the effects of climate change as many factors influence the aetiology of allergic disease, like socioeconomic status, air pollution (Saxon and Diaz-Sanchez, 2005), microbial exposures (Ege *et al.*, 2011), helminth infections (Platts-Mills *et al.*, 2006), and lifestyle (Platts-Mills, 2015), which all confound the effects of climate.

The impacts of climate change on allergenicity will probably be seen in plant-derived allergens most. Major allergens from warm-blooded animal sources or pure inorganic chemical sources are not expected to be influenced by climate change, except when climate change changes the recipient, i.e., the skin or respiratory tract. Bielory *et al.* suggested that higher temperatures for the same population increase rhinitis and asthma, i.e., change the host response (Bielory *et al.*, 2012).

The effect of climate change on pollen is not simple. Most studies show an increased pollen production and an earlier onset of the pollen season (Chapters 2 and 6; Khwarahm *et al.*, 2014; Rosenzweig *et al.*, 2008; Weber, 2012; Ziello *et al.*, 2012). Climate change will also increase pollen load and prolong allergenic pollen season due to additional pollen of invasive species like ragweed (Chapters 2, 3, and 6; Hamaoui-Laguel *et al.*, 2015; Ziska and Beggs, 2012).

Few studies show a decrease of pollen exposure, such as a trend for reduction of pollen concentrations for some pollen (González-Parrado *et al.*, 2014; Ziello *et al.*, 2012), like for grass pollen exposure in some parts of Germany (Kaminski and Glod, 2011). It should be kept in mind that pollen exposure has been determined with the Hirst-type pollen trap (Hirst, 1952) that was developed in 1952 and was a major breakthrough in pollen monitoring at the time, as it determined the concentration of pollen by sampling a known volume of air. This is a great advantage over other older routine methods that do not control sampled volume (Kishikawa *et al.*, 2009), but the method has its errors (Mullins and Emberlin, 1997). In addition, pollen concentrations depend on location of the measurement (Skjøth *et al.*, 2013). Thus, total seasonal pollen load (pollen index) is hampered by systematic errors. This is not the case for the pollen exposure parameters 'start', 'peak', and 'end' of season. These more robust parameters almost always indicate an earlier start of the pollen season due to climate change (see Chapter 6). If we then add the laboratory-controlled experiments, where higher CO_2 results in higher pollen production (Chapter 2; Rogers *et al.*, 2006), then it is safe to say that climate change prolongs the pollen season and for most species increases pollen amount (Weber, 2012). It is currently unclear whether higher temperatures or increased CO_2 concentrations are driving the allergen load of ambient air most (Ziello *et al.*, 2012).

5.2 Climate Change and Allergenicity of Pollen

The allergenicity of pollen shows a natural variability. To predict the effects of climate change, it is important to know how this variability arises. Because pollen from birch (and probably also other plant species) develop their allergen release briefly before pollination, the production of allergen is biosynthesis-dependent. This is also shown by the lack of detectable antibody staining of the allergens in unripe (about 1 week before pollination) pollen (Buters *et al.*, 2010). Pollen is released from the anthers as hydrated pollen, which then quickly, within minutes, dry in the atmosphere. The drying of pollen under normal conditions can be seen in

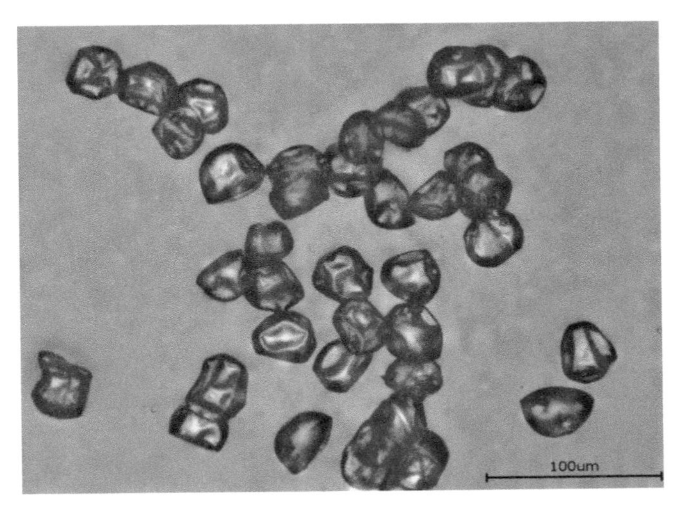

Figure 5.1. Shape of *Dactylis glomerata* pollen at ambient humidity. Pollen are emitted from the anthers as living organisms but have a limited life span. Pollen lose water within minutes and collapse, unless the exine is strong. Pollen also rehydrate quickly such as occurs for germination. Thus, the weight and probably life span of pollen depends on humidity. It is unknown whether pollen change their allergenicity upon transport through the atmosphere, but due to their dry state this is unlikely. Thus, allergenicity is determined at the source, not during transport.

Figure 5.1. Here, fresh, hydrated pollen were kept for 10 min at room temperature and room humidity. The pollen quickly lost water, exemplified by 'shrivelled' or 'collapsed' pollen. A common misconception is that pollen look like 'small footballs'. This is due to the capture and preparation technique, which utilises water-containing reagents. Upon contact with water, the pollen will hydrate, giving them their 'full', hydrated appearance. However, pollen caught from air without coming in contact with water will be dehydrated.

Biosynthesis of allergen needs active metabolism, which is unlikely when water is lacking. Thus, allergenicity of pollen depends on the biosynthesis *before* pollination. Allergenicity of pollen is determined by conditions at the location of ripening, i.e., at the source. One example of unchanged allergen release from pollen after pollination is given in Figure 5.2, which demonstrates that long-range transport of olive pollen after pollination did not change its allergenicity.

Pollen is the carrier of the allergens that induce symptoms in allergic individuals (Durham *et al.*, 2014) or lead to allergic sensitisation. In principle, it is enough to be exposed to only an allergen like a pure, recombinant allergen, to become sensitised (Pauli *et al.*, 2008). However, allergen-accompanying compounds such as the pollen-associated lipid mediators (Gilles *et al.*, 2009) or released adenosine by pollen seem to have adjuvant effects (Gilles *et al.*, 2011).

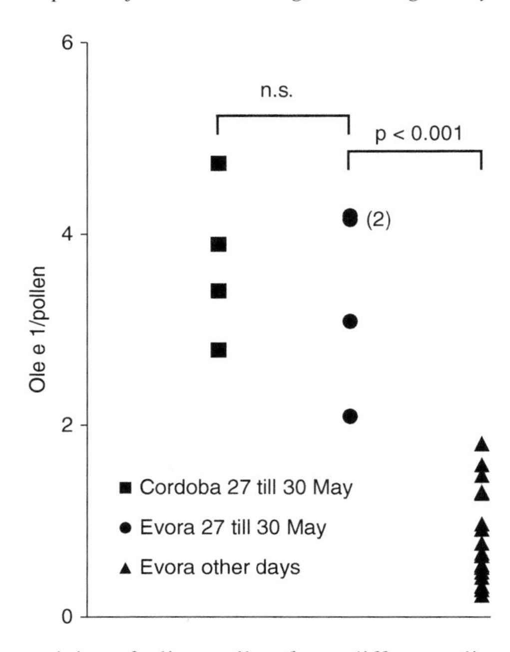

Figure 5.2. Allergenicity of olive pollen from different climatic areas. Pollen stemming from Cordoba, Spain, but caught in Evora, Portugal, in 2009 (27 till 30 May 2009) show the same allergen release potency. This shows that the inter-laboratory variability in determining allergen release capacity is low (calibration by nature), as the same pollen in Cordoba and Evora yield similar results. Then, on days when pollen comes from the surroundings of Evora, Portugal, their potency is four to five times lower than from pollen originating from Cordoba, 400 km away. Thus, olive pollen show large differences in allergen release capacity (potency), depending on the climate of their origin.

One pollen can release several antigens, like grass pollen releasing Phl p 1, Phl p 2, Phl p 4, Phl p 5, Phl p 6, Phl p 7, Phl p 11, Phl p 12, and Phl p 13 (Abou Chakra *et al.*, 2012). Although several sensitising proteins are released, sensitisations to Phl p 1, Phl p 4, and Phl p 5 seem to be the major allergens. A major allergen is that protein from a source to which >50% of the sensitised individuals react. For grass pollen, the major allergens seem to be Phl p 1 and Phl p 5 (Tripodi *et al.*, 2012) as the sensitisation to Phl p 4 is clinically not so important (Zafred *et al.*, 2013). For other sources, the major allergen is more clear like for birch, where >90% of the birch pollen–sensitised individuals react to Bet v 1, and Bet v 2 and Bet v 4 are only minor allergens (Movérare *et al.*, 2002).

The human immune system reacts to the released proteins from pollen (Pauli *et al.*, 1996, 2008). Exposure to the allergen is monitored with the proxy 'pollen'. The amount of allergen released by one pollen shows a natural variability. This natural variability was studied in the Europe-wide project HIALINE (Health Impacts of Airborne Allergen Information Network; www.hialine.eu). From HIALINE it was reported that the same amount of pollen released about ten-fold variable amounts of Bet v 1 depending on the day and location the pollen were sampled (Buters *et al.*, 2012). The same was true (twelve-fold) for olive pollen (Galan *et al.*, 2013). For grass pollen, the variability in allergen release potency per pollen was even greater, being twenty-fold (from <1 to 9 pg Phl p 5 per pollen; 5–95% confidence interval; Buters *et al.*, 2015). Pollen potency is calculated as the daily concentration of aeroallergens (pg/m^3 air) divided by daily concentration of airborne pollen (pollen grains/m^3 air).

Thus, climate change may change the number and timing of pollen in ambient air and also the potency of pollen.

Little is known about the impacts of climate change on the allergenicity of pollen. This is due to two reasons, first, that the allergenicity of pollen is seldom measured, and second, that the factors influencing the allergenicity of pollen are not understood.

In those instances where the potency of the allergen was measured, all studies found the same: like all other products of natural sources, pollen varied in the amount of allergen they released (Feo Brito *et al.*, 2011; Schäppi *et al.*, 1999a, 1999b). The next question was whether this could be influenced by climate and thus climate change.

Already Ahlholm *et al.* (1998) found that birch tree siblings grown at different climatic zones (a high- and a low-altitude garden 5 km apart with a different mean annual temperature of ~1.1°C during the growing season) changed the amount of allergen from birch pollen. Their experiments were done at a time when specific methods for allergen determination were not yet available, making the results less clear. Using a reporter assay for the pollen-specific promotor for Bet v 1, Tashpulatov *et al.* reported that this promotor became more active at increased temperatures, implying a rise in allergenicity due to temperature increases (Tashpulatov *et al.*, 2004).

HIALINE found for grass pollen (grass pollen allergy is the most important and frequent allergy against airborne pollen; Burbach *et al.*, 2009) that between two geographically different regions in Europe a consistent difference in allergen potency was detected (Figure 5.3). In Turku, Finland, and Evora, Portugal, the HIALINE project determined the amount of the major allergen Phl p 5 released from grass pollen during three consecutive years (Buters *et al.*, 2015). In all years, the Phl p 5 potency was *higher* in Finland than in Portugal. The average yearly mean temperature in the studied years was 11°C higher in Evora than in Turku. Thus indeed, climate change will change the allergenicity of pollen, but not necessarily toward more potent pollen.

We must consider that grass pollen stem from different grass species that cannot be discriminated for by conventional (microscopic) monitoring, as is currently

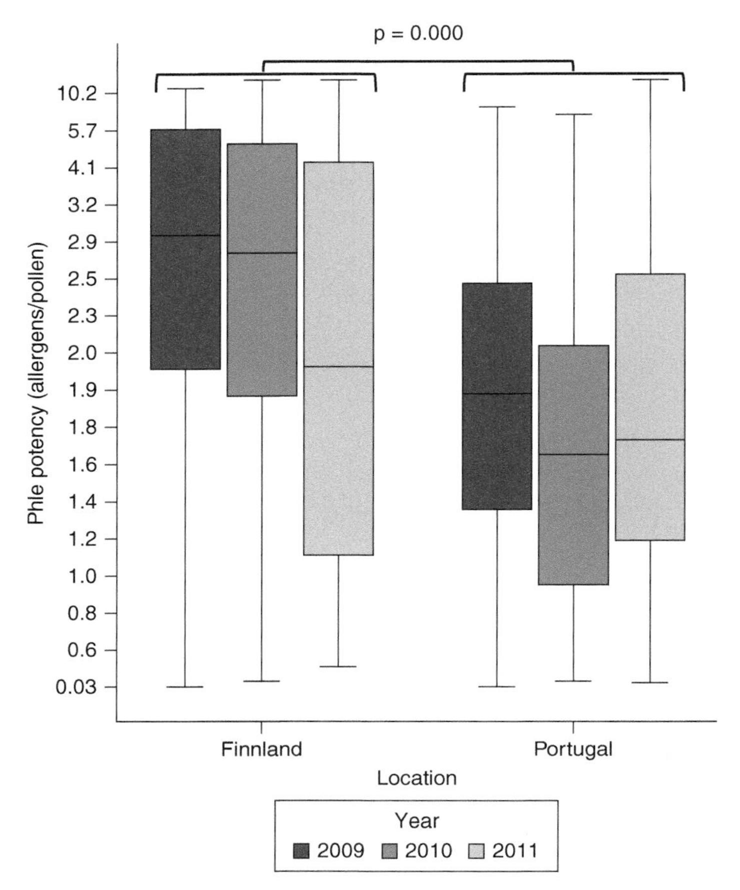

Figure 5.3. Pollen potency depending on climatic conditions. Pollen potency was determined daily during 3 years in different climatic regions. Boxplots about Phl p 5 potency of daily airborne grass pollen at two locations: Evora (Portugal) and Helsinki (Finland). Pollen potency is calculated as the daily concentration of aeroallergens (pg/m^3 air) divided by daily concentration of airborne pollen (pollen grains/m^3 air). Differences in pollen potency between both sites were tested by Mann–Whitney *U* test (nonparametric test for two independent samples). Climate change will change a local potency pattern accordingly, either by shifting to subspecies with other potencies or by varying the allergen release capacity within one species. Both effects lead to a change in allergen exposure.

the 'gold standard' in aerobiology for samples from Hirst-type pollen monitors. Hirst-type monitors are the most frequent pollen traps in the world (Buters, 2014). Therefore, the grass pollen season is a succession of populations pollinating with each grass species having potentially different pollen potency (Frenguelli *et al.*, 2010). Also, some grass species, like *Lolium perenne* and *Dactylis glomerata*, contribute more to pollen dispersion than others (León-Ruiz *et al.*, 2011).

Already, different subspecies of one plant can vary in their allergenicity as was demonstrated for the allergenicity of apples from different varieties of apple tree (Bolhaar *et al.*, 2005). It is clear that the late-blooming *Phragmitis* grass pollen has a low Phl p 5 potency, whereas other species have a higher major grass pollen allergen release potency (Ramirez *et al.*, 2010). Changes in climate might, therefore, change allergen release within one species, but could also change the succession of populations with different species and thus cause a different allergen exposure. Within one location, up to thirty different grass species can be detected (León-Ruiz *et al.*, 2011). Thus, the shift in populations by way of additional ragweed due to climate change (Hamaoui-Laguel *et al.*, 2015; Ziska *et al.*, 2011) could also take place within one species as for grass pollen, in both cases changing allergen exposure. Nevertheless, be it changed potency per pollen from the same species or changed potency due to pollen from different subspecies which changed distribution due to climate change, the exposure to allergen changes in the end.

The example of olive pollen (see Figure 5.2) shows that within one species, *Olea europaea*, pollen with different potency can be produced depending on climate. This was made clear from the case of long-range transport over about 400 km of olive pollen from Cordoba, Spain, to Evora, Portugal, in 2009 (Galan *et al.*, 2013). There, when pollen transported from Cordoba were sampled in Evora at a time when, according to phenology, no olive pollen were emitted in Evora anymore, the release of Ole e 1 (the major olive pollen allergen) was almost the same. This shows that when sampling the same pollen both stations show reproducible results (i.e., the inter-laboratory variability in determining Ole e 1 in HIALINE was negligible). However, on days when olive pollen only stemming from Portugal was sampled, their potency was five times lower. Cordoba has an average annual mean temperature ~2°C warmer than Evora. This also shows that, like with grass pollen between Finland and Portugal, different climatic conditions have an influence on pollen allergenicity.

The factors controlling allergenicity of pollen are not yet known. Besides little research available in this field, it might also be due to several factors controlling allergen potency simultaneously. For instance, the ripening of allergen in birch and probably also other pollen takes place just before pollination (see Figure 5.4). Then, when the pollen is ripe, the anthers can open and release the pollen. These two processes take place simultaneously – ripening of allergen and ripening of the anthers. The ripening of allergen takes place in the week before pollination and rises rapidly from zero allergen to high allergen levels (Buters *et al.*, 2010). During this rapid increase, the anthers control the release of the pollen. When humidity rises just before pollination, for instance due to rain, the anthers close and the pollen ripen longer. When the conditions are favourable for the anthers

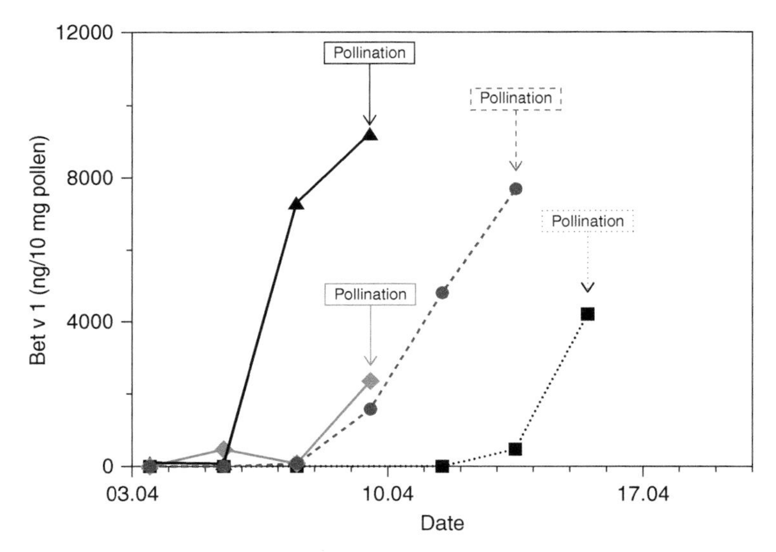

Figure 5.4. Allergen Bet v 1 expression in birch pollen upon ripening from four trees in Munich during April 2007. Pollen expresses no allergen until about 1 week before pollination. This is not due to a lack of allergen release but due to less allergen synthesis (see Buters *et al.*, 2010). Thus, variability in ripening is responsible for the variability of allergen release from pollen. Because the anthers control the release of pollen, a process that is controlled by other factors than those controlling allergen ripening, early opening of the anthers due to climate change would reduce the allergenicity of the pollen. Late opening of the anthers, for instance due to high humidity, would increase the allergenicity of pollen. Thus the effect of climate change on allergenicity is not easily predicted.

afterwards, they open and more potent pollen are emitted. From outdoor experiments, Hjelmroos *et al.* collected pollen from different sides of a white birch (*Betula pendula*) tree in Stockholm, Sweden. The north-facing branches (where temperature and light were lower) showed a decreased level of allergen expression (Hjelmroos *et al.*, 1995).

Besides temperature and humidity, CO_2 as a natural fertiliser also influences allergenicity of pollen. Several groups demonstrated in experimental settings with increased CO_2 that the allergenicity of ragweed pollen increased (El Kelish *et al.*, 2014; Singer *et al.*, 2005). The effect of increased CO_2 was not allergen-dependent, with little difference between the major and minor allergens: higher CO_2 increased allergenic proteins (El Kelish *et al.*, 2014). However, increased allergen release from ragweed pollen could not be confirmed in field experiments using natural gradients in CO_2 and temperature between urban and rural sites as a model of climate change (Ziska *et al.*, 2003). This exemplifies that several factors influence allergenicity simultaneously, and in smaller field experiments, the results of these factors might be counterintuitive. In controlled settings it is clear that elevated CO_2 increases biomass, plant constituents, and allergenicity.

Does a higher allergen exposure lead to more allergic disease? For individuals who are already allergic, a higher pollen exposure results in more allergy symptoms (Durham *et al.*, 2014; Feo Brito *et al.*, 2011). Thus a higher pollen or allergen exposure due to climate change will lead to more symptoms. Also, sensitisations increase ('becoming allergic') with exposure. In Switzerland, a community planted a robust (but allergenic) *Alnus* tree on their main road leading from the station to the centre of town. Through this road, many schoolchildren walked to school and were exposed to the pollen of this tree. Indeed, allergic sensitisations against Aln g 1, the major allergen from *Alnus* pollen, rose from low levels to much higher levels in schoolchildren from this city (Gassner *et al.*, 2013; Gehrig *et al.*, 2015). In addition, in those areas with low levels of *Ambrosia* pollen, the rate of sensitisation is lower than the neighbouring areas with higher *Ambrosia* pollen concentrations (Tosi *et al.*, 2011). Thus, also for allergic sensitisation, higher exposure leads to more allergic sensitisations. However, there is a limit to what extent higher exposure leads to more sensitisations, as very high exposures lead to allergic sensitisation but no allergic symptoms. Here a high natural exposure seems to desensitise individuals (Hesselmar *et al.*, 2001, 2003), as was also shown in controlled experiments in laboratory animals (Schwarzer *et al.*, 2011).

5.3 Thunderstorm Asthma

Climate change also results in more thunderstorms. For pollen, thunderstorms are meteorological extreme situations, and depending on the pollen species, they could rupture and release their allergenic content as small particles that could penetrate deeper into the airways and provoke more symptoms (D'Amato *et al.*, 2012). Most reports are on grass pollen that is particularly prone to rupture, but some other pollen like *Parietaria* seem to undergo this process too (D'Amato *et al.*, 2008; Schäppi *et al.*, 1997; Taylor *et al.*, 2007). More robust pollen such as that from birch trees have a more complex way to release small particles: they rupture after germination and release small allergenic particles through a broken pollen tube (Grote *et al.*, 2003; Schäppi *et al.*, 1997). Thunderstorm asthma is rare, and not many publications address this topic (Elliot *et al.*, 2014; Taylor and Jonsson, 2004; Venables *et al.*, 1997), with even less experimental evidence (Taylor *et al.*, 2002), and there is room for discussion about the causes of thunderstorm asthma (Dabrera *et al.*, 2013). Several atmospheric factors change before and during thunderstorms, and high concentrations of mould spores during thunderstorms also coincide with asthma symptoms (Grinn-Gofroń and Strzelczak, 2013). Also, already high humidity could change the amount of allergen in that fraction of the air that contains the smaller airborne particles (Buters *et al.*, 2015; Schäppi *et al.*, 1999a). As climate change increases the number of thunderstorms, the amount of rain, and humidity, this phenomenon could become more frequent. Near-fatal asthmatic attacks due to a thunderstorm during a pollen season have been reported (D'Amato *et al.*, 2013), urging that this topic needs attention of allergologists.

5.4 Climate Change and Moulds

Moulds are a major health concern in domestic and occupational settings, and contrary to pollen, legal limits for indoor mould spore concentrations are discussed (Eduard, 2009). Little is known about the allergenicity of moulds. The most predominant moulds in outdoor air are *Cladosporium* species, known for their allergens Cla h 8 and Cla c 9, and *Alternaria* species, known for their allergen Alt a 1 (www.allergome.com). *Alternaria* moulds are only a small percentage of the concentrations of *Cladosporium*, but Alt a 1, the major allergen from *Alternaria* spp., shows more sensitisations than there are against *Cladosporium* spp. (Burbach *et al.*, 2009). Some workplace mould concentrations ($>10^5$ spores/m^3) are higher that the outdoor concentrations (peak concentrations depending on the area of approximately 8,000 spores/m^3). The allergenicity of moulds is not well investigated as methods to determine mould allergens are recent and rare (Sander *et al.*, 2012). Consequently, the impact of climate change on the allergenicity of mould spores is rarely studied. Indeed, no time-series measurements of allergenicity of environmental moulds have been made. Climate change will change exposure to moulds, for instance, due to increased flooding of houses (Chapter 7; Azuma *et al.*, 2014). However, this exposure has also been determined as spores, not as allergens.

As with other temperature-sensitive organisms, a natural variability in allergenicity will exist. It is known that allergen and number of spores in moulds in a laboratory setting are influenced by temperature (Low *et al.*, 2011) or CO_2 (Lang-Yona *et al.*, 2013). For instance, *Alternaria alternata* grown on leaves grown at 600 ppm CO_2 produced three times more spores and twice the amount of allergen per plant than plants grown under current CO_2 concentrations (Wolf *et al.*, 2010).

Thus, as with pollen, increasing CO_2 will increase the number of certain moulds and their allergenicity like the *Alternaria* spp., but not all moulds (Lang-Yona *et al.*, 2013; Wolf *et al.*, 2010). Some moulds like *Aspergillus fumigatus* show an optimum growth around 360–390 ppm CO_2, with increasing allergen per spore with CO_2 increasing from pre-industrial (280 ppm) to current-day concentrations, and then decreasing allergen per spore with further increases in CO_2 (Lang-Yona *et al.*, 2013). *Cladosporium phlei* did not react to changes in CO_2 (Wolf *et al.*, 2010). The laboratory experimental results for the most allergenic mould, *Alternaria* spp., indicate a worsening of exposure: increased number of spores and allergenicity per spore due to increased temperature and CO_2.

5.5 Climate Change and Contact Allergens

The most frequent contact allergen by far is nickel (Pesonen *et al.*, 2015), eliciting about 25% of all contact allergenic cases, and regulations are in place in Europe to reduce the amount of nickel released from consumer products that come in contact with human skin to <0.2 μg Ni cm^{-2} per week. The next most frequent contact allergens are cobalt and cinnamonic acid (and its derivatives which are clinically tested as fragrance mix). Both elicit about 7% of contact allergenic cases. Nickel, cobalt, and processed products (fragrances) are independent of climate change. Climate change will unlikely change the allergenicity of the major contact allergens, but might shift the skin homeostasis (Andersen *et al.*, 2012), probably to less skin dysfunction (Byremo *et al.*, 2006).

Contact dermatitis due to poison ivy is a concern in the United States (Gladman, 2006). Mohan *et al.* (2006) could show in a 6-year, free-air CO_2 enrichment experiment that higher CO_2 resulted in more aggressive poison ivy, showing an effect of climate change on contact dermatitis. In contact dermatitis, like for pollen, the effects of climate change could be limited to plant composition changes.

5.6 Climate Change and Food Allergens

The major food allergens of concern especially in children are eggs, cow's milk, soy, wheat, and peanuts (Schmitz *et al.*, 2013). Although apples and carrots, among others, are more frequent food allergens, they are easily recognised and can be avoided, which is different from the ubiquitous 'hidden' presences of egg, cow's milk, soy, and wheat products. Peanuts need special attention, as the allergic reactions to peanuts are so severe and can be fatal, uncommon for other allergens (Du Toit *et al.*, 2015).

Eggs and cow's milk stem from warm-blooded animals that regulate their body temperature and body homeostasis independently of the environment, and it is unlikely that the amount of egg allergen (Gal d 1, egg white) in an egg will change due to climate change. Other major food allergens stem from plants, which like pollen-producing plants may react to climate change (Beggs and Walczyk, 2008). Complicating this is the commonly present natural variability in allergenicity from food products of different sources, i.e., subspecies or place of plant growth (Carnés *et al.*, 2006). Even if climate change increases the amount of allergen in certain plants, this can be offset by large differences in natural variability. Pollen are seldom commercially produced, and the natural population of pollen-producing plants will adapt slowly to changes in climate. In agricultural food production, the choice of plant subspecies and day of harvest can quickly change the allergenicity of natural products.

Because all studied examples of allergenicity from natural products varied tenfold or more in their allergen content depending on their source, the same will hold true for many other natural allergens (wasps, bees, cats, horses; Bienboire-Frosini *et al.*, 2012; Mitlehner, 2013). Because humans (the farmer), not nature, decide the plant species and time of harvest, it will be difficult to predict the allergenicity of plant food allergens and consequently the effect of climate change (Beggs, 2009). In those examples where the same plant and environment was studied, higher CO_2 increased the allergenicity of plants (Mohan *et al.*, 2006; Singer *et al.*, 2005).

References

Abou Chakra, O. R., Sutra, J. P., Thomas, E. D., *et al.* (2012). Proteomic analysis of major and minor allergens from isolated pollen cytoplasmic granules. *Journal of Proteome Research*, 11(2), 1208–1216.

Ahlholm, J. U., Helander, M. L., Savolainen, J. (1998). Genetic and environmental factors affecting the allergenicity of birch (*Betula pubescens* ssp. *czerepanovii* [Orl.] Hämet-Ahti) pollen. *Clinical and Experimental Allergy*, 28(11), 1384–1388.

Andersen, L. K., Hercogová, J., Wollina, U., Davis, M. D. P. (2012). Climate change and skin disease: a review of the English-language literature. *International Journal of Dermatology*, 51(6), 656–661.

Asher, M. I., Montefort, S., Björkstén, B., *et al.* (2006). Worldwide time trends in the prevalence of symptoms of asthma, allergic rhinoconjunctivitis, and eczema in childhood: ISAAC Phases One and Three repeat multicountry cross-sectional surveys. *The Lancet*, 368(9537), 733–743.

Azuma, K., Ikeda, K., Kagi, N., *et al.* (2014). Effects of water-damaged homes after flooding: health status of the residents and the environmental risk factors. *International Journal of Environmental Health Research*, 24(2), 158–175.

Beggs, P. J. (2009). Climate change and plant food allergens. *The Journal of Allergy and Clinical Immunology*, 123(1), 271–272.

Beggs, P. J., Walczyk, N. E. (2008). Impacts of climate change on plant food allergens: a previously unrecognized threat to human health. *Air Quality, Atmosphere & Health*, 1(2), 119–123.

Bielory, L., Lyons, K., Goldberg, R. (2012). Climate change and allergic disease. *Current Allergy and Asthma Reports*, 12(6), 485–494.

Bienboire-Frosini, C., Cozzi, A., Lafont-Lecuelle, C., *et al.* (2012). Immunological differences in the global release of the major cat allergen Fel d 1 are influenced by sex and behaviour. *The Veterinary Journal*, 193(1), 162–167.

Bolhaar, S. T. H. P., van de Weg, W. E., van Ree, R., *et al.* (2005). *In vivo* assessment with prick-to-prick testing and double-blind, placebo-controlled food challenge of allergenicity of apple cultivars. *The Journal of Allergy and Clinical Immunology*, 116(5), 1080–1086.

Burbach, G. J., Heinzerling, L. M., Edenharter, G., *et al.* (2009). GA^{2}LEN skin test study II: clinical relevance of inhalant allergen sensitizations in Europe. *Allergy*, 64(10), 1507–1515.

Buters, J. (2014). Pollen allergens and geographical factors. In: Akdis, C. A., Agache, I., eds. *Global Atlas of Allergy*. Zurich: European Academy of Allergy and Clinical Immunology, pp. 36–38.

Buters, J., Prank, M., Sofiev, M., *et al.* (2015). Variation of the group 5 grass pollen allergen content of airborne pollen in relation to geographic location and time in season. *The Journal of Allergy and Clinical Immunology*, 136(1), 87–95.

Buters, J. T. M., Thibaudon, M., Smith, M., *et al.* (2012). Release of Bet v 1 from birch pollen from 5 European countries. Results from the HIALINE study. *Atmospheric Environment*, 55, 496–505.

Buters, J. T. M., Weichenmeier, I., Ochs, S., *et al.* (2010). The allergen Bet v 1 in fractions of ambient air deviates from birch pollen counts. *Allergy*, 65(7), 850–858.

Byremo, G., Rød, G., Carlsen, K. H. (2006). Effect of climatic change in children with atopic eczema. *Allergy*, 61(12), 1403–1410.

Carnés, J., Ferrer, A., Fernández-Caldas, E. (2006). Allergenicity of 10 different apple varieties. *Annals of Allergy, Asthma & Immunology*, 96(4), 564–570.

Dabrera, G., Murray, V., Emberlin, J., *et al.* (2013). Thunderstorm asthma: an overview of the evidence base and implications for public health advice. *QJM: an International Journal of Medicine*, 106(3), 207–217.

D'Amato, G., Cecchi, L., Annesi-Maesano, I. (2012). A trans-disciplinary overview of case reports of thunderstorm-related asthma outbreaks and relapse. *European Respiratory Review*, 21(124), 82–87.

D'Amato, G., Cecchi, L., Liccardi, G. (2008). Thunderstorm-related asthma: not only grass pollen and spores. *The Journal of Allergy and Clinical Immunology*, 121(2), 537–538.

D'Amato, G., Corrado, A., Cecchi, L., *et al.* (2013). A relapse of near-fatal thunderstorm-asthma in pregnancy. *European Annals of Allergy and Clinical Immunology*, 45(3), 116–117.

Durham, S. R., Nelson, H. S., Nolte, H., *et al.* (2014). Magnitude of efficacy measurements in grass allergy immunotherapy trials is highly dependent on pollen exposure. *Allergy*, 69(5), 617–623.

Du Toit, G., Roberts, G., Sayre, P. H., *et al.* (2015). Randomized trial of peanut consumption in infants at risk for peanut allergy. *The New England Journal of Medicine*, 372(9), 803–813.

Eduard, W. (2009). Fungal spores: a critical review of the toxicological and epidemiological evidence as a basis for occupational exposure limit setting. *Critical Reviews in Toxicology*, 39(10), 799–864.

Ege, M. J., Mayer, M., Normand, A.-C., *et al.* (2011). Exposure to environmental microorganisms and childhood asthma. *The New England Journal of Medicine*, 364(8), 701–709.

El Kelish, A., Zhao, F., Heller, W., *et al.* (2014). Ragweed (*Ambrosia artemisiifolia*) pollen allergenicity: SuperSAGE transcriptomic analysis upon elevated CO_2 and drought stress. *BMC Plant Biology*, 14, 176.

Elliot, A. J., Hughes, H. E., Hughes, T. C., *et al.* (2014). The impact of thunderstorm asthma on emergency department attendances across London during July 2013. *Emergency Medicine Journal*, 31(8), 675–678.

Feo Brito, F., Gimeno, P. M., Carnés, J., *et al.* (2011). *Olea europaea* pollen counts and aeroallergen levels predict clinical symptoms in patients allergic to olive pollen. *Annals of Allergy, Asthma & Immunology*, 106(2), 146–152.

Frenguelli, G., Passalacqua, G., Bonini, S., *et al.* (2010). Bridging allergologic and botanical knowledge in seasonal allergy: a role for phenology. *Annals of Allergy, Asthma & Immunology*, 105(3), 223–227.

Galan, C., Antunes, C., Brandao, R., *et al.* (2013). Airborne olive pollen counts are not representative of exposure to the major olive allergen Ole e 1. *Allergy*, 68(6), 809–812.

Gassner, M., Gehrig, R., Schmid-Grendelmeier, P. (2013). Hay fever as a Christmas gift. *The New England Journal of Medicine*, 368(4), 393–394.

Gehrig, R., Gassner, M., Schmid-Grendelmeier, P. (2015). *Alnus* × *spaethii* pollen can cause allergies already at Christmas. *Aerobiologia*, 31(2), 239–247.

Gieras, A., Cejka, P., Blatt, K., *et al.* (2011). Mapping of conformational IgE epitopes with peptide-specific monoclonal antibodies reveals simultaneous binding of different IgE antibodies to a surface patch on the major birch pollen allergen, Bet v 1. *The Journal of Immunology*, 186(9), 5333–5344.

Gilles, S., Fekete, A., Zhang, X., *et al.* (2011). Pollen metabolome analysis reveals adenosine as a major regulator of dendritic cell-primed T_H cell responses. *The Journal of Allergy and Clinical Immunology*, 127(2), 454–461.

Gilles, S., Mariani, V., Bryce, M., *et al.* (2009). Pollen allergens do not come alone: pollen associated lipid mediators (PALMS) shift the human immune systems towards a T_H2-dominated response. *Allergy, Asthma & Clinical Immunology*, 5(1), 3.

Gladman, A. C. (2006). *Toxicodendron* dermatitis: poison ivy, oak, and sumac. *Wilderness & Environmental Medicine*, 17(2), 120–128.

González-Parrado, Z., Valencia-Barrera, R. M., Vega-Maray, A. M., Fuertes-Rodríguez, C. R., Fernández-González, D. (2014). The weak effects of climatic change on *Plantago* pollen concentration: 17 years of monitoring in Northwestern Spain. *International Journal of Biometeorology*, 58(7), 1641–1650.

Grinn-Gofroń, A., Strzelczak, A. (2013). Changes in concentration of *Alternaria* and *Cladosporium* spores during summer storms. *International Journal of Biometeorology*, 57(5), 759–768.

Grote, M., Valenta, R., Reichelt, R. (2003). Abortive pollen germination: a mechanism of allergen release in birch, alder, and hazel revealed by immunogold electron microscopy. *The Journal of Allergy and Clinical Immunology*, 111(5), 1017–1023.

Hamaoui-Laguel, L., Vautard, R., Liu, L., *et al.* (2015). Effects of climate change and seed dispersal on airborne ragweed pollen loads in Europe. *Nature Climate Change*, 5, 766–771.

Hesselmar, B., Åberg, B., Eriksson, B., Åberg, N. (2001). Allergic rhinoconjunctivitis, eczema, and sensitization in two areas with differing climates. *Pediatric Allergy and Immunology*, 12(4), 208–215.

Hesselmar, B., Åberg, B., Eriksson, B., Björkstén, B., Åberg, N. (2003). High-dose exposure to cat is associated with clinical tolerance – a modified Th2 immune response? *Clinical and Experimental Allergy*, 33(12), 1681–1685.

Hirst, J. M. (1952). An automatic volumetric spore trap. *Annals of Applied Biology*, 39(2), 257–265.

Hjelmroos, M., Schumacher, M. J., Van Hage-Hamsten, M. (1995). Heterogeneity of pollen proteins within individual *Betula pendula* trees. *International Archives of Allergy and Immunology*, 108(4), 368–376.

IPCC (2014). *Climate Change 2014: Synthesis Report. Contribution of Working Groups I, II and III to the Fifth Assessment Report of the Intergovernmental Panel on Climate Change* [Core Writing Team, Pachauri, R. K., Meyer, L. A., eds.]. Geneva, Switzerland: IPCC.

Kaminski, U., Glod, T. (2011). Are there changes in Germany regarding the start of the pollen season, the season length and the pollen concentration of the most important allergenic pollens? *Meteorologische Zeitschrift*, 20(5), 497–507.

Kelly, L. A., Erwin, E. A., Platts-Mills, T. A. E. (2012). The indoor air and asthma: the role of cat allergens. *Current Opinion in Pulmonary Medicine*, 18(1), 29–34.

Khwarahm, N., Dash, J., Atkinson, P. M., *et al.* (2014). Exploring the spatio-temporal relationship between two key aeroallergens and meteorological variables in the United Kingdom. *International Journal of Biometeorology*, 58(4), 529–545.

Kishikawa, R., Sahashi, N., Saitoh, A., *et al.* (2009). Japanese cedar airborne pollen monitoring by Durham's and Burkard samplers in Japan – estimation of the usefulness of Durham's sampler on Japanese cedar pollinosis. *Global Environmental Research*, 13(1), 55–62.

Lang-Yona, N., Levin, Y., Dannemiller, K. C., *et al.* (2013). Changes in atmospheric CO_2 influence the allergenicity of *Aspergillus fumigatus*. *Global Change Biology*, 19(8), 2381–2388.

León-Ruiz, E., Alcázar, P., Domínguez-Vilches, E., Galán, C. (2011). Study of Poaceae phenology in a Mediterranean climate. Which species contribute most to airborne pollen counts? *Aerobiologia*, 27(1), 37–50.

Low, S. Y., Dannemiller, K., Yao, M., Yamamoto, N., Peccia, J. (2011). The allergenicity of *Aspergillus fumigatus* conidia is influenced by growth temperature. *Fungal Biology*, 115(7), 625–632.

Meiler, F., Zumkehr, J., Klunker, S., *et al.* (2008). In vivo switch to IL-10-secreting T regulatory cells in high dose allergen exposure. *The Journal of Experimental Medicine*, 205(12), 2887–2898.

Mitlehner, W. (2013). Allergy against horses. Are curly horses an alternative for horse-allergic riders? *Allergo Journal*, 22(4), 244–251.

Mohan, J. E., Ziska, L. H., Schlesinger, W. H., *et al.* (2006). Biomass and toxicity responses of poison ivy (*Toxicodendron radicans*) to elevated atmospheric CO_2. *Proceedings of the National Academy of Sciences of the United States of America*, 103(24), 9086–9089.

Movérare, R., Westritschnig, K., Svensson, M., *et al.* (2002). Different IgE reactivity profiles in birch pollen-sensitive patients from six European populations revealed by recombinant allergens: an imprint of local sensitization. *International Archives of Allergy and Immunology*, 128(4), 325–335.

Mullins, J., Emberlin, J. (1997). Sampling pollens. *Journal of Aerosol Science*, 28(3), 365–370.

Pauli, G., Larsen, T. H., Rak, S., *et al.* (2008). Efficacy of recombinant birch pollen vaccine for the treatment of birch-allergic rhinoconjunctivitis. *The Journal of Allergy and Clinical Immunology*, 122(5), 951–960.

Pauli, G., Oster, J. P., Deviller, P., *et al.* (1996). Skin testing with recombinant allergens rBet v 1 and birch profilin, rBet v 2: diagnostic value for birch pollen and associated allergies. *The Journal of Allergy and Clinical Immunology*, 97(5), 1100–1109.

Pesonen, M., Jolanki, R., Filon, F. L., *et al.* (2015). Patch test results of the European baseline series among patients with occupational contact dermatitis across Europe – analyses of the European Surveillance System on Contact Allergy network, 2002–2010. *Contact Dermatitis*, 72(3), 154–163.

Platts-Mills, T. A. E. (2015). The allergy epidemics: 1870–2010. *The Journal of Allergy and Clinical Immunology*, 136(1), 3–13.

Platts-Mills, T. A. E., Erwin, E. A., Woodfolk, J. A., Heymann, P. W. (2006). Environmental factors influencing allergy and asthma. In: Crameri, R. ed. *Allergy and Asthma in Modern Society: A Scientific Approach*. Chemical Immunology and Allergy, 91. Basel: Karger, pp. 3–15.

Ramirez, J.-M., Brembilla, N. C., Sorg, O., *et al.* (2010). Activation of the aryl hydrocarbon receptor reveals distinct requirements for IL-22 and IL-17 production by human T helper cells. *European Journal of Immunology*, 40(9), 2450–2459.

Rogers, C. A., Wayne, P. M., Macklin, E. A., *et al.* (2006). Interaction of the onset of spring and elevated atmospheric CO_2 on ragweed (*Ambrosia artemisiifolia* L.) pollen production. *Environmental Health Perspectives*, 114(6), 865–869.

Rosenzweig, C., Karoly, D., Vicarelli, M., *et al.* (2008). Attributing physical and biological impacts to anthropogenic climate change. *Nature*, 453(7193), 353–357.

Sander, I., Zahradnik, E., van Kampen, V., *et al.* (2012). Development and application of mold antigen-specific enzyme-linked immunosorbent assays (ELISA) to quantify airborne antigen exposure. *Journal of Toxicology and Environmental Health, Part A: Current Issues*, 75(19–20), 1185–1193.

Saxon, A., Diaz-Sanchez, D. (2005). Air pollution and allergy: you are what you breathe. *Nature Immunology*, 6(3), 223–226.

Schäppi, G. F., Taylor, P. E., Pain, M. C. F., *et al.* (1999a). Concentrations of major grass group 5 allergens in pollen grains and atmospheric particles: implications for hay fever and allergic asthma sufferers sensitized to grass pollen allergens. *Clinical and Experimental Allergy*, 29(5), 633–641.

Schäppi, G. F., Taylor, P. E., Staff, I. A., Rolland, J. M., Suphioglu, C. (1999b). Immunologic significance of respirable atmospheric starch granules containing major birch allergen Bet v 1. *Allergy*, 54(5), 478–483.

Schäppi, G. F., Taylor, P. E., Staff, I. A., Suphioglu, C., Knox, R. B. (1997). Source of Bet v 1 loaded inhalable particles from birch revealed. *Sexual Plant Reproduction*, 10(6), 315–323.

Schmitz, R., Ellert, U., Kalcklösch, M., Dahm, S., Thamm, M. (2013). Patterns of sensitization to inhalant and food allergens – findings from the German Health Interview and Examination Survey for Children and Adolescents. *International Archives of Allergy and Immunology*, 162(3), 263–270.

Schwarzer, M., Repa, A., Daniel, C., *et al.* (2011). Neonatal colonization of mice with *Lactobacillus plantarum* producing the aeroallergen Bet v 1 biases towards Th1 and T-regulatory responses upon systemic sensitization. *Allergy*, 66(3), 368–375.

Singer, B. D., Ziska, L. H., Frenz, D. A., Gebhard, D. E., Straka, J. G. (2005). Increasing Amb a 1 content in common ragweed (*Ambrosia artemisiifolia*) pollen as a function of rising atmospheric CO_2 concentration. *Functional Plant Biology*, 32(7), 667–670.

Skjøth, C. A., Ørby, P. V., Becker, T., *et al.* (2013). Identifying urban sources as cause of elevated grass pollen concentrations using GIS and remote sensing. *Biogeosciences*, 10(1), 541–554.

Tashpulatov, A. S., Clement, P., Akimcheva, S. A., *et al.* (2004). A model system to study the environment-dependent expression of the *Bet v 1a* gene encoding the major birch pollen allergen. *International Archives of Allergy and Immunology*, 134(1), 1–9.

Taylor, P. E., Flagan, R. C., Valenta, R., Glovsky, M. M. (2002). Release of allergens as respirable aerosols: a link between grass pollen and asthma. *The Journal of Allergy and Clinical Immunology*, 109(1), 51–56.

Taylor, P. E., Jacobson, K. W., House, J. M., Glovsky, M. M. (2007). Links between pollen, atopy and the asthma epidemic. *International Archives of Allergy and Immunology*, 144(2), 162–170.

Taylor, P. E., Jonsson, H. (2004). Thunderstorm asthma. *Current Allergy and Asthma Reports*, 4(5), 409–413.

Tosi, A., Wüthrich, B., Bonini, M., Pietragalla-Köhler, B. (2011). Time lag between Ambrosia sensitisation and Ambrosia allergy: a 20-year study (1989–2008) in Legnano, northern Italy. *Swiss Medical Weekly*, 141, w13253.

Traidl-Hoffmann, C., Jakob, T., Behrendt, H. (2009). Determinants of allergenicity. *The Journal of Allergy and Clinical Immunology*, 123(3), 558–566.

Tripodi, S., Frediani, T., Lucarelli, S., *et al.* (2012). Molecular profiles of IgE to *Phleum pratense* in children with grass pollen allergy: implications for specific immunotherapy. *The Journal of Allergy and Clinical Immunology*, 129(3), 834–839.

Tschopp, J. M., Sistek, D., Schindler, C., *et al.* (1998). Current allergic asthma and rhinitis: diagnostic efficiency of three commonly used atopic markers (IgE, skin prick tests, and Phadiatop®). Results from 8329 randomized adults from the SAPALDIA study. *Allergy*, 53(6), 608–613.

Venables, K. M., Allitt, U., Collier, C. G., *et al.* (1997). Thunderstorm-related asthma – the epidemic of 24/25 June 1994. *Clinical and Experimental Allergy*, 27(7), 725–736.

Weber, R. W. (2012). Impact of climate change on aeroallergens. *Annals of Allergy, Asthma & Immunology*, 108(5), 294–299.

Willumsen, N., Holm, J., Christensen, L. H., Würtzen, P. A., Lund, K. (2012). The complexity of allergic patients' IgE repertoire correlates with serum concentration of allergen-specific IgE. *Clinical & Experimental Allergy*, 42(8), 1227–1236.

Wolf, J., O'Neill, N. R., Rogers, C. A., Muilenberg, M. L., Ziska, L. H. (2010). Elevated atmospheric carbon dioxide concentrations amplify *Alternaria alternata* sporulation and total antigen production. *Environmental Health Perspectives*, 118(9), 1223–1228.

Zafred, D., Nandy, A., Pump, L., Kahlert, H., Keller, W. (2013). Crystal structure and immunologic characterization of the major grass pollen allergen Phl p 4. *The Journal of Allergy and Clinical Immunology*, 132(3), 696–703.

Ziello, C., Sparks, T. H., Estrella, N., *et al.* (2012). Changes to airborne pollen counts across Europe. *PLoS One*, 7(4), e34076.

Ziska, L. H., Beggs, P. J. (2012). Anthropogenic climate change and allergen exposure: the role of plant biology. *The Journal of Allergy and Clinical Immunology*, 129(1), 27–32.

Ziska, L. H., Gebhard, D. E., Frenz, D. A., *et al.* (2003). Cities as harbingers of climate change: common ragweed, urbanization, and public health. *The Journal of Allergy and Clinical Immunology*, 111(2), 290–295.

Ziska, L., Knowlton, K., Rogers, C., *et al.* (2011). Recent warming by latitude associated with increased length of ragweed pollen season in central North America. *Proceedings of the National Academy of Sciences of the United States of America*, 108(10), 4248–4251.

6

Impacts of Climate Change on Allergen Seasonality

LEWIS H. ZISKA
Crop Systems and Global Change
Agricultural Research Service
US Department of Agriculture

6.1 Introduction

Projected changes in population and the concomitant needs related to food and fuel will continue to drive greenhouse gas emissions for the foreseeable future. Since 1970, the concentration of global atmospheric carbon dioxide (CO_2) has risen by 22%, from ~325 parts per million (ppm) to 395 ppm, with projected estimates of increase to 600–800 ppm by the end of the century (Meinshausen *et al.*, 2011).

The physics of greenhouse gas emissions and the resulting impacts on climate stability have been well reviewed by the scientific community (see Chapter 1). Less well characterised, however, are the consequences of these ongoing and projected changes with respect to the nexus of plant biology and human health. Such interactions can be varied and reflect everything from toxicology to human nutrition (Ziska *et al.*, 2009). Among these potential linkages, the influence of climate on pollen production, aeroallergens, and allergic disease is widely recognised and has generated significant interest by both the general public and the scientific community (Cecchi *et al.*, 2010; Reid and Gamble, 2009; Shea *et al.*, 2008). In examining potential interactions, it is important to elucidate those specific aspects of anthropogenic climate change that are likely to alter the production, distribution, and dispersion of aeroallergens and the organisms that produce them, i.e., trees, grasses, weeds, and fungus.

One of the most recognised and fundamental aspect is related to global surface temperatures. Although the term 'global warming' infers an average increase in temperature, in reality, differential increases in global surface temperatures are occurring. Such differential increases are manifest in the relative accumulation of distinct greenhouse gases. For example, in the tropics, where it is warm and humid, water vapour is the dominant greenhouse gas, and rising atmospheric CO_2 will exert a smaller effect on surface temperatures than in regions with less water vapour (e.g., poles). The Intergovernmental Panel on

Climate Change (IPCC) assessments have, in fact, recognised this variation in warming and emphasised that land surface temperatures will increase more than the global average, as will temperatures at high latitudes and elevations (Hansen *et al.*, 2006; IPCC, 2013). Similarly, winters should warm faster than summers because winters are colder (and hence drier) with a greater relative response of surface temperatures to rising CO_2 levels. Such regional and seasonal changes in temperature associated with anthropogenic climate change have obvious biological consequences in regard to flowering phenology and aeroallergens. Indeed, shifts in phenology are among the most consistent findings in studies of global climate change (Cleland *et al.*, 2007; IPCC, 2014; Sparks *et al.*, 2000). Clearly, such changes in the timing and duration of flowering and pollen production can, and will, lead to seasonal changes in human exposure to aeroallergens with subsequent alterations in sensitisation and potential exacerbation of allergic diseases.

A second, less recognised link is that CO_2, in addition to being a greenhouse gas, is the sole supplier of carbon for photosynthesis and growth. As the bulk (~95%) of all plant species (including all tree species) lack optimal amounts of CO_2, the rapid increase in this gas globally represents the availability of an essential resource and will have significant consequences for plant physiology. In general, the effect of rising CO_2 concentrations is to also accelerate (i.e., advance) flowering time; however, the extent to which flowering phenology is affected is species-dependent (Springer and Ward, 2007).

Overall, there is widespread empirical evidence of an association between anthropogenic environmental change, aerobiology, and allergic disease, and it is likely that projected changes in temperature and CO_2 concentration will intensify the public health consequences of such a link. The current chapter is an attempt to assimilate and assess what is known regarding the impact of anthropogenic factors, particularly temperature and CO_2 changes, on a key aspect of allergic disease – changes in the timing and length of aeroallergen production from known biological sources.

6.2 Biological Sources of Aeroallergens

In general, it is recognised that among plant species the greatest annual production of aeroallergens occurs from tree pollen, followed by weed and grass pollen (Bielory *et al.*, 2012). Relevant plant species are anemophilous, or wind-pollinated. Because such pollination is less reliable than insect pollination (entomophily), such plants produce large amounts of pollen to ensure fertilisation and are acknowledged aeroallergen sources.

Aeroallergen exposure is strongly correlated with allergic rhinitis (hay fever). Allergic diseases, including hay fever, impact approximately one-third of the population in the United States and are associated with significant health-care costs and lost productivity. A study by the American Academy of Allergy, Asthma, and Immunology (AAAAI, 2015) reported that 54.6% of people in the United States test positive for an allergic response to one or more allergens, and more than 34 million Americans have been diagnosed with asthma. Blando *et al.* (2012) report a clear link between ambient pollen levels and hospital-based health-care utilisation in Australia, Canada, Spain, and the United States. The AAAAI estimates that about 36 million persons in the United States have seasonal ragweed allergies (Centers for Disease Control and Prevention, 2014).

Fungi, also commonly called moulds, comprise a taxonomic kingdom with many members ubiquitous in nature. Many fungi produce distinct sexual and asexual airborne spores. Three classes of primary concern to the allergist are the Zygomycetes (e.g., Mucor, Rhizopus), the Ascomycetes (e.g., yeast, powdery mildews), and the Basidiomycetes (e.g., rusts, smuts; Burge, 1985). As with pollen, exposure to fungal spores is associated with allergy and asthma symptoms (Institute of Medicine, 2004; Salo *et al.*, 2006); however, the specifics of the relationship are not completely understood (Portnoy *et al.*, 2008).

Alternaria alternata is a common fungus with over 300 plant hosts and is a known source of allergenic spores. Among patients with asthma from six regions of the world, 11.9%, on average, were sensitised to *A. alternata*, with the proportion as high as 28.2% in Portland, Oregon; in addition, sensitivity to *A. alternata* was more prevalent among patients with more severe asthma (Zureik *et al.*, 2002).

6.3 Temperature and Pollen Seasonality

6.3.1 Spring

It is generally recognised that seasonal pollen occurs in three temporal phases – trees in spring, grasses in summer, and weeds, especially ragweed (*Ambrosia* species) in autumn. To date, there are a number of phenological studies showing a clear link between anthropogenic climate change, warming winter and spring temperatures, and earlier flowering times in a wide variety of tree species that are known sources of allergenic pollen (Table 6.1). There is considerable variation in these studies that reflects the time period examined and regional differences in temperature, etc.; however, for all tree species examined, flowering is now occurring, on average, approximately 2 weeks earlier than it did relative to the mid-twentieth-century temperature average. These differences are, with few exceptions, statistically significant (Table 6.1). Although clinical associations with seasonal variation

Table 6.1. *Changes in initiation (start) dates (as day of the year) for pollen release for known allergenic species of trees in response to recent warming trends.*

Species	Country	Time period	Start	Start	Difference	Reference
Ash	Netherlands	1970s–1990s	92	88	−3	van Vliet *et al.*, 2002
Birch	Belgium	1982–2000	102	84	−18*	Emberlin *et al.*, 2002
Birch	Finland	1975–2004	130	118	−12*	Yli-Panula *et al.*, 2009
Birch	Netherlands	1970s–1990s	106	94	−10*	van Vliet *et al.*, 2002
Birch	Switzerland	1982–2000	105	85	−20*	Emberlin *et al.*, 2002
Elder	Netherlands	1970s–1990s	169	154	−15***	van Vliet *et al.*, 2002
Elm	Netherlands	1970s–1990s	79	57	−22**	van Vliet *et al.*, 2002
Juniper	Netherlands	1970s–1990s	71	51	−20**	van Vliet *et al.*, 2002
Oak	Netherlands	1970s–1990s	135	117	−18***	van Vliet *et al.*, 2002
Oak	Spain	1970s–1990s	89	78	−11	García-Mozo *et al.*, 2006
Olive	Italy	1981–2007	143	107	−35*	Ariano *et al.*, 2010
Olive	Spain	1990s–2000s	98	90	−8*	Galán *et al.*, 2005
Pine	Netherlands	1970s–1990s	144	135	−9	van Vliet *et al.*, 2002
Poplar	Netherlands	1970s–1990s	84	66	−18**	van Vliet *et al.*, 2002
Willow	Netherlands	1970s–1990s	82	70	−12*	van Vliet *et al.*, 2002

Species are listed alphabetically by common name. Statistically significant differences in start date between the start and the end of the time period are indicated with $^*p<0.05$, $^{**}p<0.01$, and $^{***}p<0.001$.

are rare; warmer and more humid weather trends associated with earlier springs have been associated with aggravation of exercise-induced asthma during the birch (*Betula* spp.) pollen season (Dapul-Hidalgo and Bielory, 2012).

Projections of these observations to further warming are unclear. For oak, earlier flowering associated with warming winter and spring temperatures has been observed over a 50-year period (García-Mozo *et al.*, 2006). Projected changes in temperature associated with a doubling of CO_2 concentration by the end of the twenty-first century would, in turn, result in significant changes in both earlier pollen initiation and concentration (30 days and 50% higher; García-Mozo *et al.*, 2006); whether there is a temporal limit on how early flowering can occur is not yet clear.

Whether current trends in earlier flowering and pollen release continue is complicated by other aspects of tree biology. Floral initiation in boreal hardwoods may require a minimal cumulative amount of chilling temperatures (vernalisation) followed by accumulation of heat in order to break dormancy. For example, in birch, a ubiquitous source of spring aeroallergens, different species exhibit different chilling requirements (Miller-Rushing and Primack, 2008). Theoretically, if winters warmed sufficiently such that chilling requirements were not met, no floral initiation would occur. Therefore, the seasonal rate and occurrence of warming temperatures (e.g., winter versus spring) could result in earlier or later anthesis and pollen release depending on the tree species–specific need for vernalisation. A study of spring and winter flowering trends among plant communities observed trends in the onset of flowering depended strongly on the magnitude of a given species response to autumn/winter versus spring warming (Cook *et al.*, 2012).

An additional climatic factor for some tree species is the phenomenon of masting, or synchronous flowering/fruit production in species such as oak and beech. Environmental parameters associated with climate change such as drought and summer temperatures (in addition to spring/winter temperatures) can all alter masting, in part by increasing the overall carbohydrate concentration within the tree (Piovesan and Adams, 2001; Sork *et al.*, 1993). Increases in carbohydrate can, in turn, 'prime' oak for floral induction (Piovesan and Adams, 2001). Masting in turn would result in earlier and above-average pollen production, but is not sustainable over years; that is, a masting year is often followed by a year of poor floral production.

In evaluating the role of anthropogenic forcing in tree pollen seasonality, it is also important to consider range and distribution (Chapter 3). Warming alone is likely to have significant effects on the redistribution of tree species. Some tree species may shift poleward; others, depending on the rate of warming, may not adapt quickly enough to migrate in response to temperatures and become limited in range (e.g., Cheaib *et al.*, 2012; Iverson and Prasad, 1998). Such longer-term variation is likely to be species-dependent, but may also reflect ecological or biome changes that reflect future disease and pest pressures that could limit tree distribution (e.g., temperature-induced changes in oak disease; Bergot *et al.*, 2004). Overall, these changes, in addition to biological drivers at the single tree level, will greatly influence pollen aerobiology by altering species exposure and pollen release.

Table 6.2. *Changes in initiation (start) dates (as day of the year) for pollen release for known allergenic species of weeds and grasses in response to recent warming trends.*

Species	Country	Time period	Start	Start	Difference	Reference
Dock	Netherlands	1970s–1990s	141	136	−5	van Vliet *et al.*, 2002
Goosefoot	Netherlands	1970s–1990s	199	187	−12*	van Vliet *et al.*, 2002
Grasses	Italy	1981–2007	134	128	−6	Ariano *et al.*, 2010
Grasses	Netherlands	1970s–1990s	129	118	−11	van Vliet *et al.*, 2002
Mugwort	Netherlands	1970s–1990s	220	207	−12*	van Vliet *et al.*, 2002
Mugwort	Poland	1995–2004	203	196	−7	Stach *et al.*, 2007
Nettle	Netherlands	1970s–1990s	158	147	−11**	van Vliet *et al.*, 2002

Species are listed alphabetically by common name. Statistically significant differences in start date between the start and the end of the time period are indicated with $*p<0.05$ and $**p<0.01$.

6.3.2 Summer

Because of the relative effects of CO_2 and water vapour, anthropogenic impacts on temperature and seasonality should be more evident at the beginning and end of the growing season (i.e., spring and autumn). Hence, the relative effect of warming should be less for plant aeroallergens during the summer. A compilation of weed and grass species that are known allergenic pollen producers during the summer indicated no significant effect in relation to earlier pollen release for dock in the Netherlands, grasses in Italy and the Netherlands, and mugwort in Poland (Table 6.2). However, earlier flowering and pollen release was observed for goosefoot (*Chenopodia album*), as well as mugwort and nettle in the Netherlands (Table 6.2). Overall, the effect of increasing temperatures on earlier anthesis was less evident for summer pollen producers.

For future projected temperature increases, it is unclear if additional warming will alter temporal pollen production and release for weed and grass species during the summer. It has been shown in crop species that retardation of floral biology, particularly induced pollen sterility, occurs at much lower temperatures than for other plant organs (e.g., leaf growth; Hatfield *et al.*, 2011). For example, in maize, the optimal temperature for pollen release is 18–22°C, and pollen production ceases at 35°C (Hatfield *et al.*, 2011). Whether similar affects occur for anemophilous weeds is unknown; such adverse effects of high temperature extremes would be manifest, however, in altered changes in the seasonality and production of summer weeds and grasses.

As with tree distribution, additional warming is likely to alter the range of known allergenic weeds and grasses (Chapter 3). Some of these species, in addition to being sources of summer aeroallergens, are also agronomic weeds, and their range is projected to shift towards the poles with additional warming (e.g., McDonald *et al.*, 2009). Because many of these weed species are annuals, it is likely that their ability to migrate biogeographically will be greater, and therefore migration will occur at a faster rate than that of tree species. Hence, seasonality and temporal exposure of plant-based allergens (e.g., trees in the spring, weeds/grasses in the summer) will likely be altered as growing seasons shift poleward. Such shifts in floristic zones with temperature are already evident in the hardiness zone maps of North America (Arbor Day Foundation, 2006).

6.3.3 Autumn

For plant-based allergens that occur in the autumn, the impact of anthropogenic warming is likely to be evident in pollen season duration, rather than earlier anthesis. Onset of flowering for these species is determined by shorter light periods following the summer solstice, and pollen production is ongoing until the initial frost. Although there are a number of plant species that produce pollen in autumn, among the most prolific and most allergenic is ragweed (*Ambrosia* species; Arbes Jr *et al.*, 2005). *Ambrosia artemisiifolia* (common or short ragweed) is among the most common causes of respiratory allergies in North America. The pollen is tricolporate, with a spiny, granular surface which aids in surface adhesion. *A. artemisiifolia* can grow extensively in disturbed sites, and a single plant can release about one billion pollen grains in a single season (Oswalt and Marshall Jr, 2008).

Simulated warming was shown to increase pollen production and diameter in western ragweed (*A. psilostachya*) in situ as well as extend the length of the growing season, but the effects on pollen seasonality and temporal shedding were not examined (Wan *et al.*, 2002, 2005). In 2011, in cooperation with the National Allergy Bureau in the United States and the Aerobiology Research Laboratories in Canada, ragweed pollen season duration was examined in North America (Ziska *et al.*, 2011). A synthesis of these data along a 2,200-km north-south latitudinal transect demonstrated that the duration of the ragweed pollen season has been increasing since 1995 as a function of latitude. These latitudinal changes in turn, were associated with a delay in the first autumn frost and a lengthening of the frost-free period (Ziska *et al.*, 2011). The length of ragweed pollen exposure increased with increasing latitude, with the season extended by up to 27 days at latitudes above ~44°N (Ziska *et al.*, 2011). This potential northward expansion of ragweed allergy season with anthropogenic warming in North America is reflected in similar migratory concerns for northern Europe (Dahl *et al.*, 1999).

If current projections of season length and poleward range expansion continue with additional warming, it is likely that new populations will encounter ragweed pollen. As with summer weeds and grasses, a number of ragweed species are annuals and are highly adapted to physical disturbance and environmental extremes (Ziska *et al.*, 2007) and, therefore, likely to experience a more rapid range expansion than tree species. Whether the southern limit of ragweed distribution in the Northern Hemisphere will also shift poleward with additional warming, however, is unknown.

In published studies to date, there is a clear trend in pollen seasonality associated with anthropogenic warming with a clear association between rising temperature and the start of the pollen season for tree species. This finding is consistent with a number of phenological studies indicating changes in other organismal cues, including bud break. Change in pollen seasonality is less evident (but not absent) for summer grasses and weeds; evidence is mixed for grass pollens, with studies showing regional trend differences in the UK (Emberlin, 1994) and earlier start dates in Switzerland (Clot, 2003), but no effect for grasses in Italy or the Netherlands (Table 6.2). Overall trends for a delay in the end of the pollen season among autumn allergens are unclear (Fitter and Fitter, 2002), although more recent assessments suggest an extension of the vegetative growing season in autumn over mid- and high latitudes in North America (Zhu *et al.*, 2012). However, floral development per se is more likely to be limited by temperature early in the growing season with low ambient temperatures relative to later in the season when temperatures exceed thermal optimums. Hence, differential effects of warming temperatures on flowering times and pollen seasonality would be expected. However, for ragweed (*Ambrosia* species), cumulative data across a north-south transect was correlated with a delay in the onset of the first frost in recent decades, suggesting an overall increase in the pollen season length for this genus. For all plant-based aeroallergens, further changes in seasonality will be dependent on the rate of temperature increase, the degree of species adaptation, and potential changes in range and distribution. Unfortunately, with few exceptions, little is known regarding flowering and pollen seasonality associated with temperature and migratory patterns.

6.4 Atmospheric CO_2 and Pollen Seasonality

Even in the absence of any anthropogenic change in surface temperatures, increases in atmospheric CO_2 concentration can directly affect key aspects of floral biology, including flowering times. The effect of rising CO_2, however, is variable, with some plant species showing shorter times to anthesis, whereas others show delays in flowering or no effect (Johnston and Reekie, 2008; Springer and Ward, 2007).

The biological basis for the direct effect of CO_2 on phenology and development for allergenic plants (or plants in general) has yet to be entirely elucidated. Elevated CO_2 has been shown to influence the expression of floral initiation genes in *Arabidopsis thaliana* (Springer *et al.*, 2008). Because it stimulates photosynthesis and growth rates, increasing CO_2 may also reduce the time required to reach the minimal critical size for floral induction (He and Bazzaz, 2003). In addition, increases in CO_2 concentration may increase leaf carbohydrate concentration, with additional assimilate available for floral initiation, size, or number (Ward and Strain, 1999).

To accommodate their size, experimental data on the biological role of rising CO_2 on tree floral phenology are often obtained from large (>100 m^2) Free-Air CO_2 Enrichment (FACE) facilities. Research on loblolly pine (*Pinus taeda*) at the Duke University FACE location demonstrated that elevated CO_2 alone could induce earlier (and greater) seasonal pollen production (LaDeau and Clark, 2001).

For weedy and grass species, elevated CO_2 has been shown to stimulate the production of male and female flowers in *Silene latifolia*, a summer annual, with the extent of stimulation much greater in male flowers (Wang, 2005). Other summer species, including a number of allergenic plants, demonstrate a wide range of responses; for example, flowering in pigweed (*Amaranthus retroflexus*) is accelerated (Garbutt *et al.*, 1990) as it is in annual bluegrass (Leishman *et al.*, 1999), whereas flowering in other grass species may be delayed or unchanged (Jablonski *et al.*, 2002; Springer and Ward, 2007). For common ragweed, no effect of recent (Ziska and Caulfield, 2000) or projected increases in atmospheric CO_2 (700 ppm; Garbutt *et al.*, 1990; Rogers *et al.*, 2006; Ziska and Caulfield, 2000) was observed for flowering times.

6.5 Temperature, CO_2, and Pollen Seasonality

While it is clear that each environmental factor – CO_2 and temperature – can influence flowering biology and the seasonality of aeroallergen production, it is likely that both factors will change in parallel (IPCC, 2013). How then will the interaction of these aspects alter floral phenology? Johnston and Reekie (2008) examined a subset of Asteraceae species and showed that the relative effect of rising CO_2 and temperature is greater for early- rather than late-blooming species, with no effect of rising temperature and CO_2 on plant species that were day-neutral. For allergenic plants, this would suggest a greater synergy for tree species (late winter, spring) rather than weeds and grasses in the summer.

Unfortunately, from a methodological point of view, it is difficult to physically encompass mature, floral-bearing trees as a means to regulate both CO_2 and temperature (FACE systems can enhance CO_2, but not temperature, over a large area). To overcome such methodological difficulties, researchers have been studying urban climates as a potential surrogate for future climatic conditions (e.g., George *et al.*, 2007). As urbanisation already reflects localised increases in CO_2, air temperatures, and growing season length, it may serve as an experimental means to examine potential changes in flowering phenology (Neil and Wu, 2006). For example, in a 5-year study of CO_2 and temperature across an urban–rural transect for Baltimore, Maryland, the recorded increases in atmospheric CO_2, air temperature, and frost-free days were consistent with short-term projections of climate change (George *et al.*, 2007; IPCC, 2013; Unger, 1999; Ziska *et al.*, 2004). Analysis of such transects is subject to complicating microenvironmental perturbations, e.g., nitrogen deposition, and does not allow separation of CO_2 from temperature effects; in addition, the relative urban–rural differences in these abiotic variables will be population-dependent (i.e., larger cities will have, in general, a larger temperature increase relative to their surroundings). However, at present, such transects, while imperfect, do provide a means to assess concurrent increases in CO_2 and temperature over large surface areas in situ (George *et al.*, 2007; Neil and Wu, 2006).

Roetzer *et al.* (2000) analysed data for four spring-flowering shrubs and trees from ten central European observation stations between 1951 and 1995 and noted that, despite regional differences, species flowered earlier in urban relative to rural areas by 2–4 days. Rodríguez-Rajo *et al.* (2010) observed that a city in Poland (Eskulap) showed earlier release and later conclusion of the pollen season (and overall longer season) in urban versus rural/suburban locales. They noted that while pollen samplers are often located in urban areas, the data on pollen seasonality are often extended into surrounding rural populations and suggested that extrapolation of these data may be in error. Spieksma *et al.* (2003) analysed pollen grain counts for several plant species in five cities, with data on birch showing an increasing seasonal trend at all sites. Bryce *et al.* (2010) also observed that urbanisation could alter the allergenicity of birch (*Betula*) pollen (Chapter 5).

A glasshouse experiment with common ragweed (*A. artemisiifolia*) where the onset of spring and CO_2 were manipulated found that simulation of earlier springs resulted in increased pollen production at ambient CO_2, and that elevated CO_2 increased pollen production only in middle- and late-blooming ragweed plants (Rogers *et al.*, 2006). In an urban–rural transect for Baltimore, Maryland, with the urban site averaging ~2°C higher temperature and ~20% more CO_2 relative to the rural site, ragweed grew faster, flowered earlier, and produced significantly more pollen per plant than the rural site (Ziska *et al.*, 2003; Figure 6.1).

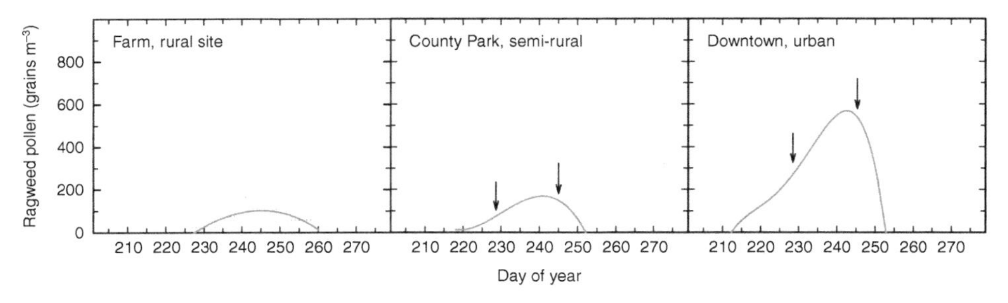

Figure 6.1. Changes in ragweed pollen seasonality as a function of urbanisation along a rural–urban transect for Baltimore, Maryland, USA. Arrows indicate the start and maximal pollen times for the rural site (control). Data are from Ziska *et al.* (2003).

6.6 Fungi and Aeroallergen Seasonality

Aeroallergens are generated by both plants and fungi (mould is a term that refers to hundreds of saprophytic fungi species). Daily increases in outdoor aeroallergens were associated with concomitant increases in hospital admissions for asthma, with fungal spores having a greater influence than pollens for a given proportionate increase (Dales *et al.*, 2004). Fungal species, including *Alternaria*, *Cladosporium*, and *Aspergillus*, are associated in some studies with higher prevalence of asthma hospital admissions (Cecchi *et al.*, 2010). Fungi, in general, require high humidity, moisture, and temperatures above a minimal physiological threshold. Hence, anthropogenic climate change, specifically warmer temperatures, more frost-free days, and changes in precipitation (e.g., humidity), is almost certain to have significant impacts on fungal life cycles, including sporulation (Peden and Reed, 2010). In addition, extreme weather events (e.g., thunderstorms, flooding) could also favour an increase in spore counts (Celenza *et al.*, 1996). Overall, earlier springs and associated changes in the hydrological cycle may, in turn, have an effect on the seasonality of fungal growth, sporulation, and the timing of aeroallergen exposure; however, these interactions have not been well studied.

As plants are often hosts for fungi, it is likely that there will be indirect consequences of rising CO_2 levels in addition to direct climatic effects on seasonality of spore production and release. For example, *A. alternata* is widely recognised as a significant aeroallergen. Longer growing seasons would extend agricultural harvests with later exposure to *Alternaria* spores. Studies examining increases in local temperature and agricultural cereal production for two towns in the United Kingdom indicated that *Alternaria* spore counts increased from the early 1970s to the late 1990s with the spore season starting earlier and increasing in duration (Corden and Millington, 2001; Corden *et al.*, 2003). For quaking aspen (*Populus tremuloides*), fungal propagules including *A. alternata* increased in leaf litter at twice ambient CO_2 concentrations (Klironomos *et al.*, 1997). However, a later study by Kelly *et al.* (2010) did not show any significant change in microbial colonisation of *P. tremuloides, Salix alba*, or *Acer saccharum* leaf litter at 720 ppm CO_2. For timothy grass (*Phleum pratense*) inoculated with *A. alternata* and grown at 300 ppm (preindustrial), 400 ppm (current), or 500 or 600 ppm (future) CO_2 concentration, *A. alternata* produced nearly three times the number of spores and more than twice the total antigenic protein per plant at 600 ppm CO_2 relative to the 300 ppm CO_2 (Wolf *et al.*, 2010). While the role of CO_2 on fungal-produced aeroallergens and seasonality in plant systems has not been entirely elucidated, many fungi/moulds are associated with plant decomposition, so that CO_2-induced increases in plant growth, combined with warmer temperatures and longer growing season duration, are likely to increase fungal spore load and, potentially, lengthen exposure times.

6.7 Localised Anthropogenic Change and Aeroallergen Seasonality

In addition to the rapid rise in atmospheric CO_2 and surface temperatures, humans are also significantly changing the extent of tropospheric ozone (e.g., Stocker *et al.*, 2013); land use patterns, especially soil disturbance (Pielke Sr *et al.*, 2002); and extreme meteorological events. Such changes, in addition to those in temperature and CO_2, are also likely to impact plant biology, including flowering and aeroallergen production times and duration.

6.7.1 Air Pollutants

Ozone is generated by photochemical dissociation of nitrogen dioxide and is temperature-dependent; consequently, it is likely to be exacerbated with climate change (Alexander *et al.*, 2013). There is little question that air pollution, notably nitrous oxide and ozone, can directly exacerbate asthma (Peden and Reed, 2010); however, less is known regarding the interaction of ozone on aeroallergen production and seasonality. For cypress (*Cupressus arizonica*), higher pollen allergenicity was observed in polluted relative to less polluted cities (Suárez-Cervera *et al.*, 2008); similarly, for rye (*Secale cereale*), elevated ozone (80 ppb) increased pollen allergenicity (Eckl-Dorna *et al.*, 2010). Timothy grass (*P. pratense*) also released more allergens when exposed to supra-ambient concentrations of nitrogen dioxide and ozone (Motta *et al.*, 2006). In contrast to these findings, elevated ozone had no effect on ragweed growth or anthesis (Ziska, 2002) and no effect on pollen allergenicity (Kanter *et al.*, 2013).

In a seminal study, Knowlton *et al.* (2007) mapped ragweed pollen prevalence with regional locales with at least one ozone exceedance day per year. Interestingly, fourteen of the fifteen urban sites included in the Asthma and Allergy Foundation of America as among 'the most challenging places to live with asthma' were located in areas that had overlapping ozone and ragweed risks. Additional information as to whether ozone and aeroallergens act in a synergistic or antagonistic fashion, however, is lacking (Reid and Gamble, 2009); however, with warming springs and hotter summer temperatures, potential overlap between pollen release and higher ozone concentrations should be examined in greater detail.

Overall, the interrelationships between climate, CO_2, and air pollution in regard to duration and seasonality of aeroallergens are complex. It seems probable in reviewing what is known that findings of a given study reflect specific aspects such as pollen species, the kind of air pollutant, urbanisation, etc. (D'Amato and Cecchi, 2008).

6.7.2 Land Use and Disturbance

Although climate change is of obvious concern, other anthropogenic changes related to land cover and invasive species deserve additional consideration. Deforestation, for example, has obvious consequences for spring pollen production for nearby human populations. Yet, it is also clear than in ecological terms, deforestation, or major disturbances of existing vegetation either through deliberate land use change or increased storm intensity, will alter the demography of plant communities (Bellard *et al.*, 2012). Shifts in plant populations (Chapter 3), in turn, will also alter aeroallergen seasonality and exposure times.

Human-induced disturbances are evident in the distribution and persistence of common ragweed. In a study of urban ragweed populations, Ziska *et al.* (2007) noted that while CO_2 and temperature increases associated with urbanisation did result in earlier flowering (and greater pollen release) in ragweed, populations over time were likely to be dependent on soil disturbance at regular intervals. For example, the introduction and expansion of ragweed in eastern Europe (where it is considered an invasive species) is associated with major socioeconomic transitions that increase the area of disturbed land (Bassett and Crompton, 1975; Csontos *et al.*, 2010). Following the collapse of communism in 1989, agricultural fields belonging to cooperatives were divided into smaller parcels with a greater emphasis on mechanisation, particularly tillage and soil disturbance for weed management. Soil disturbance, in turn, was a likely factor in the establishment and spread of ragweed (Bazzaz, 1970).

6.7.3 Extreme Meteorological Events

Greater occurrence of severe weather events, such as hurricanes, thunderstorms, etc., is also likely to influence the temporal distribution of aeroallergens (Reid and Gamble, 2009). As most pollen is produced from anemophilous plant species, changes in windstorm frequency and intensity could result in greater propagation and spread of such species. Thunderstorm winds and precipitation can also degrade pollen grains with a greater release of allergen particles (D'Amato *et al.*, 2007). A 6-year study found that fungal spore aeroallergen concentration was also associated with a significant increase in asthma rates on thunderstorm versus non-thunderstorm days relative to pollen or air pollutants (Dales *et al.*, 2003). Hurricanes, such as Katrina in 2005, impact allergic diseases (Ratard *et al.*, 2006) with flooding-induced allergy-causing moulds occurring both indoors and outdoors and higher mould spore counts in October and November 2005 in New Orleans (Solomon *et al.*, 2006; see also Chapter 7). While individual extreme weather events associated with anthropogenic climate change cannot always be distinguished from their natural happenstance, greater temporal frequency of such events has obvious consequences for seasonality and duration of aeroallergen exposure (Beggs, 2004).

6.8 Experimental Unknowns

There are numerous knowledge gaps and research challenges associated with anthropogenic climate change and respiratory health. Many of these will only be apparent as climate change progresses; others, however, are becoming self-evident. Among the latter is the issue of changing surface temperatures, rising CO_2, and changes in the seasonality and duration of outdoor aeroallergens. Yet, it is also evident that there are a number of key research needs where additional information is needed to both assess vulnerability and determine adaptation strategies. The following is meant to be illustrative of these needs, not a complete and exhaustive assessment of them.

6.8.1 Methodology

Among the most frustrating aspects of determining temporal aeroallergen trends is the differential use of pollen-monitoring equipment and the inability to quantitatively compare pollen and spore numbers across regions. At present, most US counting stations use Burkard Samplers, but other volumetric devices (e.g., Rotorod Sampler) and gravimetric methods (e.g., Durham Sampler) have also been used in recent decades. Unfortunately, relative comparisons between these sampling methods are not possible (Frenz, 1999). Current development of global long-term (multidecadal) pollen/spore data sets and temporal and quantitative spatial analysis of aeroallergens has been hampered as a result. A consistent, uniform set of methodologies is urgently needed to ensure consistency and compatibility of aeroallergen information across spatial and temporal scales.

6.8.2 Monitoring

Methods are needed to improve aeroallergen detection and increase the number of monitoring sites regionally and globally. In addition, there is an obvious need to assimilate existing long-term data series on all aeroallergens and to quantify recent trends using urbanisation, seasons, and phenological cues as dependent factors. Efforts should also be made to link successional changes in plant communities associated with climate change to monitoring information.

6.8.3 Assessments

There is a critical need to link health, climate, and ecological data (e.g., geographic information systems) to understand the biological and physical links between moulds and allergic diseases in relation to meteorological variables and to assess how urban warming or land use changes may interact or act as a surrogate for projected climate change. Additional experimental and field studies are needed to both confirm and expand our knowledge base on how existing and future CO_2 and temperature may alter aeroallergen production, distribution, and duration. Studies are also needed in this regard for other anthropogenic aspects such as air pollution, extreme weather events and land use, as well as long-term ecological trends such as succession and evolutionary adaptation.

6.8.4 Surveillance and Epidemiology

Established levels of allergen exposure that determine risk and vulnerability for asthma and rhinitis as a function of seasonality are needed. Long-term data sets can be used to link these levels to weather and climate parameters as a means to begin model development for assessing risk of temporal changes in aeroallergen exposure among different populations. These data, in turn, would be critical in developing and improving the means to project changes in aeroallergen ranges, distribution, and concentration in association with both global and regional climate. Further, this information needs to be linked to health-care professionals as a means to develop and implement plans to minimise aeroallergen exposure and health threats. Finally, where applicable, there is a critical need to initiate, develop, and implement more effective, sustainable approaches for aeroallergen control (e.g., ragweed management).

6.9 Conclusions

Evidence to date indicates that anthropogenic changes in climate and atmospheric CO_2 are likely to alter the timing and distribution of plant- and fungal-based aeroallergens with potential effects on the severity and prevalence of allergic diseases in the human population. There is also limited evidence that other anthropogenic factors such as land use, urbanisation, air pollution, etc., may exacerbate these trends. There is a clear need to continue and expand research not only with respect to underlying links between climate, CO_2, and aeroallergen seasonality but also to the degree to which human actions can manage changes in vegetation and aeroallergen exposure. Overall, it is hoped that the current synthesis will act as an impetus for additional resources directed towards a better understanding of climate change, CO_2, and aeroallergen seasonality. Such resources will be critical to derive time-relevant scientific and policy solutions that can minimise public health impacts.

Acknowledgements

The comments and editing suggestions of Dr Mark Schwartz from the University of Wisconsin are gratefully appreciated.

References

Alexander, L. V., Allen, S. K., Bindoff, N. L., *et al.* (2013). Summary for policymakers. In: Stocker, T. F., Qin, D., Plattner, G.-K., *et al.*, eds. *Climate Change 2013: The Physical Science Basis. Contribution of Working Group I to the Fifth Assessment Report of the Intergovernmental Panel on Climate Change.* Cambridge, UK and New York, NY, USA: Cambridge University Press, pp. 3–29.

American Academy of Allergy, Asthma, and Immunology (AAAAI) (2015). Allergy Statistics. Available at: www.aaaai.org/about-the-aaaai/newsroom/allergy-statistics .aspx. Accessed 19 June 2015.

Arbes Jr, S. J., Gergen, P. J., Elliott, L., Zeldin, D. C. (2005). Prevalences of positive skin test responses to 10 common allergens in the US population: results from the Third National Health and Nutrition Examination Survey. *The Journal of Allergy and Clinical Immunology*, 116(2), 377–383.

Arbor Day Foundation (2006). 2006 Hardiness Zone Map. Available at: www.arborday .org/media/zones.cfm. Accessed 19 June 2015.

Ariano, R., Canonica, G. W., Passalacqua, G. (2010). Possible role of climate changes in variations in pollen seasons and allergic sensitizations during 27 years. *Annals of Allergy, Asthma & Immunology*, 104(3), 215–222.

Bassett, I. J., Crompton, C. W. (1975). The biology of Canadian weeds. 11. *Ambrosia artemisiifolia* L. and *A. psilostachya* DC. *Canadian Journal of Plant Science*, 55(2), 463–476.

Bazzaz, F. A. (1970). Secondary dormancy in the seeds of the common ragweed *Ambrosia artemisiifolia*. *Bulletin of the Torrey Botanical Club*, 97(5), 302–305.

Beggs, P. J. (2004). Impacts of climate change on aeroallergens: past and future. *Clinical and Experimental Allergy*, 34(10), 1507–1513.

Bellard, C., Bertelsmeier, C., Leadley, P., Thuiller, W., Courchamp, F. (2012). Impacts of climate change on the future of biodiversity. *Ecology Letters*, 15(4), 365–377.

Bergot, M., Cloppet, E., Pérarnaud, V., *et al.* (2004). Simulation of potential range expansion of oak disease caused by *Phytophthora cinnamomi* under climate change. *Global Change Biology*, 10(9), 1539–1552.

Bielory, L., Lyons, K., Goldberg, R. (2012). Climate change and allergenic disease. *Current Allergy and Asthma Reports*, 12(6), 485–494.

Blando, J., Bielory, L., Nguyen, V., Diaz, R., Jeng, H. A. (2012). Anthropogenic climate change and allergic diseases. *Atmosphere*, 3(1), 200–212.

Bryce, M., Drews, O., Schenk, M. F., *et al.* (2010). Impact of urbanization on the proteome of birch pollen and its chemotactic activity on human granulocytes. *International Archives of Allergy and Immunology*, 151(1), 46–55.

Burge, H. A. (1985). Fungus allergens. *Clinical Reviews in Allergy*, 3(3), 319–329.

Cecchi, L., D'Amato, G., Ayres, J. G., *et al.* (2010). Projections of the effects of climate change on allergic asthma: the contribution of aerobiology. *Allergy*, 65(9), 1073–1081.

Celenza, A., Fothergill, J., Kupek, E., Shaw, R. J. (1996). Thunderstorm associated asthma: a detailed analysis of environmental factors. *British Medical Journal*, 312(7031), 604–607.

Centers for Disease Control and Prevention (2014). Allergies and Hay Fever. Available at: www.cdc.gov/nchs/fastats/allergies.htm. Accessed 22 June 2015.

Cheaib, A., Badeau, V., Boe, J., *et al.* (2012). Climate change impacts on tree ranges: model intercomparison facilitates understanding and quantification of uncertainty. *Ecology Letters*, 15(6), 533–544.

Cleland, E. E., Chuine, I., Menzel, A., Mooney, H. A., Schwartz, M. D. (2007). Shifting plant phenology in response to global change. *Trends in Ecology and Evolution*, 22(7), 357–365.

Clot, B. (2003). Trends in airborne pollen: an overview of 21 years of data in Neuchâtel (Switzerland). *Aerobiologia*, 19(3–4), 227–234.

Cook, B. I., Wolkovich, E. M., Parmesan, C. (2012). Divergent responses to spring and winter warming drive community level flowering trends. *Proceedings of the National Academy of Sciences of the United States of America*, 109(23), 9000–9005.

Corden, J. M., Millington, W. M. (2001). The long-term trends and seasonal variation of the aeroallergen *Alternaria* in Derby, UK. *Aerobiologia*, 17(2), 127–136.

Corden, J. M., Millington, W. M., Mullins, J. (2003). Long-term trends and regional variation in the aeroallergen *Alternaria* in Cardiff and Derby UK – are differences in climate and cereal production having an effect? *Aerobiologia*, 19(3–4), 191–199.

Csontos, P., Vitalos, M., Barina, Z., Kiss, L. (2010). Early distribution and spread of *Ambrosia artemisiifolia* in Central and Eastern Europe. *Botanica Helvetica*, 120(1), 75–78.

Dahl, Å., Strandhede, S.-O., Wihl, J.-Å. (1999). Ragweed – an allergy risk in Sweden? *Aerobiologia*, 15(4), 293–297.

Dales, R. E., Cakmak, S., Judek, S., *et al.* (2003). The role of fungal spores in thunderstorm asthma. *Chest*, 123(3), 745–750.

Dales, R. E., Cakmak, S., Judek, S., *et al.* (2004). Influence of outdoor aeroallergens on hospitalization for asthma in Canada. *The Journal of Allergy and Clinical Immunology*, 113(2), 303–306.

D'Amato, G., Cecchi, L. (2008). Effects of climate change on environmental factors in respiratory allergic diseases. *Clinical and Experimental Allergy*, 38(8), 1264–1274.

D'Amato, G., Liccardi, G., Frenguelli, G. (2007). Thunderstorm-asthma and pollen allergy. *Allergy*, 62(1), 11–16.

Dapul-Hidalgo, G., Bielory, L. (2012). Climate change and allergic diseases. *Annals of Allergy, Asthma & Immunology*, 109(3), 166–172.

Eckl-Dorna, J., Klein, B., Reichenauer, T. G., Niederberger, V., Valenta, R. (2010). Exposure of rye (*Secale cereale*) cultivars to elevated ozone levels increases the allergen content in pollen. *The Journal of Allergy and Clinical Immunology*, 126(6), 1315–1317.

Emberlin, J. (1994). The effects of patterns in climate and pollen abundance on allergy. *Allergy*, 49(s18), 15–20.

Emberlin, J., Detandt, M., Gehrig, R., *et al.* (2002). Responses in the start of *Betula* (birch) pollen seasons to recent changes in spring temperatures across Europe. *International Journal of Biometeorology*, 46(4), 159–170. See also erratum (2003). 47(2), 113–115.

Fitter, A. H., Fitter, R. S. R. (2002). Rapid changes in flowering time in British plants. *Science*, 296(5573), 1689–1691.

Frenz, D. A. (1999). Comparing pollen and spore counts collected with the Rotorod Sampler and Burkard spore trap. *Annals of Allergy, Asthma & Immunology*, 83(5), 341–349.

Galán, C., García-Mozo, H., Vázquez, L., *et al.* (2005). Heat requirement for the onset of the *Olea europaea* L. pollen season in several sites in Andalusia and the effect of the expected future climate change. *International Journal of Biometeorology*, 49(3), 184–188.

Garbutt, K., Williams, W. E., Bazzaz, F. A. (1990). Analysis of the differential response of five annuals to elevated CO_2 during growth. *Ecology*, 71(3), 1185–1194.

García-Mozo, H., Galán, C., Jato, V., *et al.* (2006). *Quercus* pollen season dynamics in the Iberian Peninsula: response to meteorological parameters and possible consequences of climate change. *Annals of Agriculture and Environmental Medicine*, 13(2), 209–224.

George, K., Ziska, L. H., Bunce, J. A., Quebedeaux, B. (2007). Elevated atmospheric CO_2 concentration and temperature across an urban-rural transect. *Atmospheric Environment*, 41(35), 7654–7665.

Hansen, J., Sato, M., Ruedy, R., *et al.* (2006). Global temperature change. *Proceedings of the National Academy of Sciences of the United States of America*, 103(39), 14288–14293.

Hatfield, J. L., Boote, K. J., Kimball, B. A., *et al.* (2011). Climate impacts on agriculture: implications for crop production. *Agronomy Journal*, 103(2), 351–370.

He, J.-S., Bazzaz, F. A. (2003). Density-dependent responses of reproductive allocation to elevated atmospheric CO_2 in *Phytolacca americana*. *New Phytologist*, 157(2), 229–239.

Institute of Medicine (US) (2004). Committee on Damp Indoor Spaces and Health. *Damp Indoor Spaces and Health*. Washington, DC: The National Academies Press.

IPCC (2013). *Climate Change 2013: The Physical Science Basis. Contribution of Working Group I to the Fifth Assessment Report of the Intergovernmental Panel on Climate Change* [Stocker, T. F., Qin, D., Plattner, G.-K., *et al.*, eds.]. Cambridge, UK and New York, NY: Cambridge University Press.

IPCC (2014). *Climate Change 2014: Impacts, Adaptation, and Vulnerability. Part A: Global and Sectoral Aspects. Contribution of Working Group II to the Fifth Assessment Report of the Intergovernmental Panel on Climate Change* [Field, C. B., Barros, V. R., Dokken, D. J., *et al.*, eds.]. Cambridge, UK and New York, NY: Cambridge University Press.

Iverson, L. R., Prasad, A. M. (1998). Predicting abundance of 80 tree species following climate change in the eastern United States. *Ecological Monographs*, 68(4), 465–485.

Jablonski, L. M., Wang, X., Curtis, P. S. (2002). Plant reproduction under elevated CO_2 conditions: a meta-analysis of reports on 79 crop and wild species. *New Phytologist*, 156(1), 9–26.

Johnston, A., Reekie, E. (2008). Regardless of whether rising atmospheric carbon dioxide levels increase air temperature, flowering phenology will be affected. *International Journal of Plant Sciences*, 169(9), 1210–1218.

Kanter, U., Heller, W., Durner, J., *et al.* (2013). Molecular and immunological characterization of ragweed (*Ambrosia artemisiifolia* L.) pollen after exposure of the plants to elevated ozone over a whole growing season. *PLoS One*, 8(4), e61518.

Kelly, J. J., Bansal, A., Winkelman, J., *et al.* (2010). Alteration of microbial communities colonizing leaf litter in a temperate woodland stream by growth of trees under conditions of elevated atmospheric CO_2. *Applied and Environmental Microbiology*, 76(15), 4950–4959.

Klironomos, J. N., Rillig, M. C., Allen, M. F., *et al.* (1997). Increased levels of airborne fungal spores in response to *Populus tremuloides* grown under elevated atmospheric CO_2. *Canadian Journal of Botany*, 75(10), 1670–1673.

Knowlton, K., Rotkin-Ellman, M., Solomon, G. (2007). *Sneezing and Wheezing: How Global Warming Could Increase Ragweed Allergies, Air Pollution, and Asthma.* New York: Natural Resources Defense Council. Available at: www.nrdc.org/globalwarming/sneezing/sneezing.pdf. Accessed 23 June 2015.

LaDeau, S. L., Clark, J. S. (2001). Rising CO_2 levels and the fecundity of forest trees. *Science*, 292(5514), 95–98.

Leishman, M. R., Sanbrooke, K. J., Woodfin, R. M. (1999). The effects of elevated CO_2 and light environment on growth and reproductive performance of four annual species. *New Phytologist*, 144(3), 455–462.

McDonald, A., Riha, S., DiTommaso, A., DeGaetano, A. (2009). Climate change and the geography of weed damage: analysis of U.S. maize systems suggests the potential for significant range transformations. *Agriculture, Ecosystems & Environment*, 130(3–4), 131–140.

Meinshausen, M., Smith, S. J., Calvin, K., *et al.* (2011). The RCP greenhouse gas concentrations and their extensions from 1765 to 2300. *Climatic Change*, 109(1–2), 213–241.

Miller-Rushing, A. J., Primack, R. B. (2008). Effects of winter temperatures on two birch (*Betula*) species. *Tree Physiology*, 28(4), 659–664.

Motta, A. C., Marliere, M., Peltre, G., Sterenberg, P. A., Lacroix, G. (2006). Traffic-related air pollutants induce the release of allergen-containing cytoplasmic granules from grass pollen. *International Archives of Allergy and Immunology*, 139(4), 294–298.

Neil, K., Wu, J. (2006). Effects of urbanization on plant flowering phenology: a review. *Urban Ecosystems*, 9(3), 243–257.

Oswalt, M. L., Marshall Jr, G. D. (2008). Ragweed as an example of worldwide allergen expansion. *Allergy, Asthma, and Clinical Immunology*, 4(3), 130–135.

Peden, D., Reed, C. E. (2010). Environmental and occupational allergies. *The Journal of Allergy and Clinical Immunology*, 125(2), S150–S160.

Pielke Sr, R. A., Marland, G., Betts, R. A., *et al.* (2002). The influence of land-use change and landscape dynamics on the climate system: relevance to climate-change policy beyond the radiative effect of greenhouse gases. *Philosophical Transactions of the Royal Society of London. Series A: Mathematical, Physical and Engineering Sciences*, 360(1797), 1705–1719.

Piovesan, G., Adams, J. M. (2001). Masting behaviour in beech: linking reproduction and climatic variation. *Canadian Journal of Botany*, 79(9), 1039–1047.

Portnoy, J. M., Barnes, C. S., Kennedy, K. (2008). Importance of mold allergy in asthma. *Current Allergy and Asthma Reports*, 8(1), 71–78.

Ratard, R., Brown, C. M., Ferdinands, J., *et al.* (2006). Health concerns associated with mold in water-damaged homes after Hurricanes Katrina and Rita – New Orleans Area, Louisiana, October 2005. *Morbidity and Mortality Weekly Report*, 55(2), 41–44.

Reid, C. E., Gamble, J. L. (2009). Aeroallergens, allergic disease, and climate change: impacts and adaptation. *EcoHealth*, 6(3), 458–470.

Rodríguez-Rajo, F. J., Fdez-Sevilla, D., Stach, A., Jato, V. (2010). Assessment between pollen seasons in areas with different urbanization level related to local vegetation sources and differences in allergen exposure. *Aerobiologia*, 26(1), 1–14.

Roetzer, T., Wittenzeller, M., Haeckel, H., Nekovar, J. (2000). Phenology in central Europe – differences and trends of spring phenophases in urban and rural areas. *International Journal of Biometeorology*, 44(2), 60–66.

Rogers, C. A., Wayne, P. M., Macklin, E. A., *et al.* (2006). Interaction of the onset of spring and elevated atmospheric CO_2 on ragweed (*Ambrosia artemisiifolia* L.) pollen production. *Environmental Health Perspectives*, 114(6), 865–869.

Salo, P. M., Arbes Jr, S. J., Sever, M., *et al.* (2006). Exposure to *Alternaria alternata* in US homes is associated with asthma symptoms. *The Journal of Allergy and Clinical Immunology*, 118(4), 892–898.

Shea, K. M., Truckner, R. T., Weber, R. W., Peden, D. B. (2008). Climate change and allergic disease. *The Journal of Allergy and Clinical Immunology*, 122(3), 443–453.

Solomon, G. M., Hjelmroos-Koski, M., Rotkin-Ellman, M., Hammond, S. K. (2006). Airborne mold and endotoxin concentrations in New Orleans, Louisiana, after flooding, October through November 2005. *Environmental Health Perspectives*, 114(9), 1381–1386.

Sork, V. L., Bramble, J., Sexton, O. (1993). Ecology of mast-fruiting in three species of North American deciduous oaks. *Ecology*, 74(2), 528–541.

Sparks, T. H., Jeffree, E. P., Jeffree, C. E. (2000). An examination of the relationship between flowering times and temperature at the national scale using long-term phenological records from the UK. *International Journal of Biometeorology*, 44(2), 82–87.

Spieksma, F. Th. M., Corden, J. M., Detandt, M., *et al.* (2003). Quantitative trends in annual totals of five common airborne pollen types (*Betula*, *Quercus*, Poaceae, *Urtica*, and *Artemisia*), at five pollen-monitoring stations in western Europe. *Aerobiologia*, 19(3–4), 171–184.

Springer, C. J., Orozco, R. A., Kelly, J. K., Ward, J. K. (2008). Elevated CO_2 influences the expression of floral-initiation genes in *Arabidopsis thaliana*. *New Phytologist*, 178(1), 63–67.

Springer, C. J., Ward, J. K. (2007). Flowering time and elevated atmospheric CO_2. *New Phytologist*, 176(2), 243–255.

Stach, A., García-Mozo, H., Prieto-Baena, J. C., *et al.* (2007). Prevalence of *Artemisia* species pollinosis in western Poland: impact of climate change on aerobiological trends, 1995–2004. *Journal of Investigational Allergology and Clinical Immunology*, 17(1), 39–47.

Stocker, T. F., Qin, D., Plattner, G.-K., *et al.* (2013). Technical summary. In: Stocker, T. F., Qin, D., Plattner, G.-K., *et al.*, eds. *Climate Change 2013: The Physical Science Basis. Contribution of Working Group I to the Fifth Assessment Report of the Intergovernmental Panel on Climate Change*. Cambridge, UK and New York, NY, USA: Cambridge University Press, pp. 33–115.

Suárez-Cervera, M., Castells, T., Vega-Maray, A., *et al.* (2008). Effects of air pollution on Cup a 3 allergen in *Cupressus arizonica* pollen grains. *Annals of Allergy, Asthma & Immunology*, 101(1), 57–66.

Unger, J. (1999). Comparisons of urban and rural bioclimatological conditions in the case of a Central-European city. *International Journal of Biometeorology*, 43(3), 139–144.

van Vliet, A. J. H., Overeem, A., De Groot, R. S., Jacobs, A. F. G., Spieksma, F. T. M. (2002). The influence of temperature and climate change on the timing of pollen release in the Netherlands. *International Journal of Climatology*, 22(14), 1757–1767.

Wan, S., Hui, D., Wallace, L., Luo, Y. (2005). Direct and indirect effects of experimental warming on ecosystem carbon processes in a tallgrass prairie. *Global Biogeochemical Cycles*, 19(2), GB2014.

Wan, S., Yuan, T., Bowdish, S., *et al.* (2002). Response of an allergenic species, *Ambrosia psilostachya* (Asteraceae), to experimental warming and clipping: implications for public health. *American Journal of Botany*, 89(11), 1843–1846.

Wang, X. (2005). Reproduction and progeny of *Silene latifolia* (Caryophyllaceae) as affected by atmospheric CO_2 concentration. *American Journal of Botany*, 92(5), 826–832.

Ward, J. K., Strain, B. R. (1999). Elevated CO_2 studies: past, present and future. *Tree Physiology*, 19(4–5), 211–220.

Wolf, J., O'Neill, N. R., Rogers, C. A., Muilenberg, M. L., Ziska, L. H. (2010). Elevated atmospheric carbon dioxide concentrations amplify *Alternaria alternata* sporulation and total antigen production. *Environmental Health Perspectives*, 118(9), 1223–1228.

Yli-Panula, E., Fekedulegn, D. B., Green, B. J., Ranta, H. (2009). Analysis of airborne *Betula* pollen in Finland; a 31-year perspective. *International Journal of Environmental Research and Public Health*, 6(6), 1706–1723.

Zhu, W., Tian, H., Xu, X., *et al.* (2012). Extension of the growing season due to delayed autumn over mid and high latitudes in North America during 1982–2006. *Global Ecology and Biogeography*, 21(2), 260–271.

Ziska, L. H. (2002). Sensitivity of ragweed (*Ambrosia artemisiifolia*) growth to urban ozone concentrations. *Functional Plant Biology*, 29(11), 1365–1369.

Ziska, L. H., Bunce, J. A., Goins, E. W. (2004). Characterization of an urban-rural CO_2/temperature gradient and associated changes in initial plant productivity during secondary succession. *Oecologia*, 139(3), 454–458.

Ziska, L. H., Caulfield, F. A. (2000). Rising CO_2 and pollen production of common ragweed (*Ambrosia artemisiifolia*), a known allergy-inducing species: implications for public health. *Australian Journal of Plant Physiology*, 27(10), 893–898.

Ziska, L. H., Epstein, P. R., Schlesinger, W. H. (2009). Rising CO_2, climate change, and public health: exploring the links to plant biology. *Environmental Health Perspectives*, 117(2), 155–158.

Ziska, L. H., Gebhard, D. E., Frenz, D. A., *et al.* (2003). Cities as harbingers of climate change: common ragweed, urbanization, and public health. *The Journal of Allergy and Clinical Immunology*, 111(2), 290–295.

Ziska, L. H., George, K., Frenz, D. A. (2007). Establishment and persistence of common ragweed (*Ambrosia artemisiifolia* L.) in disturbed soil as a function of an urban-rural macro-environment. *Global Change Biology*, 13(1), 266–274.

Ziska, L., Knowlton, K., Rogers, C., *et al.* (2011). Recent warming by latitude associated with increased length of ragweed pollen season in central North America. *Proceedings of the National Academy of Sciences of the United States of America*, 108(10), 4248–4251.

Zureik, M., Neukirch, C., Leynaert, B., *et al.* (2002). Sensitisation to airborne moulds and severity of asthma: cross sectional study from European Community respiratory health survey. *British Medical Journal*, 325(7361), 411–418.

7

Impacts of Climate Change on Indoor Allergens

GINGER L. CHEW AND SHUBHAYU SAHA
Centers for Disease Control and Prevention
National Center for Environmental Health
Division of Environmental Hazards and Health Effects
Air Pollution and Respiratory Health Branch

7.1 Introduction

Although both genetic and environmental factors are important in explaining the large variations in asthma within and between populations (Asher *et al.*, 1995), environmental factors are likely to offer the greatest opportunities for change. We have learned from past and ongoing studies of exposures in the urban environment (and more recently the suburban and rural environments) that environmental factors are inextricably combined with social factors (e.g., poverty, substandard housing) which we do not completely understand (Chew *et al.*, 2006a; Gold and Acevedo-Garcia, 2005; Gruchalla *et al.*, 2005; Kitch *et al.*, 2000; Lin *et al.*, 2012; Matsui *et al.*, 2003, 2004; Perry *et al.*, 2012; Phipatanakul *et al.*, 2005). As such, it is difficult to definitively attribute an increased burden of allergy to climate change. However, it appears that climate change–related factors (both direct and indirect) could be associated with a changing profile of indoor allergens and in some cases either an increase or a decrease in indoor allergen concentrations in buildings, most notably the home environment.

7.2 Dust Mite Allergens

Dust mite allergen is one of the most ubiquitous indoor allergens known (Platts-Mills *et al.*, 1997). Much of the allergenicity attributed to dust mites is due to their faecal pellets, measuring from 10–40 μm in diameter (Tovey *et al.*, 1981). Dust mites are microscopic and feed on human skin scales, fungi, and other forms of organic matter (Colloff, 2009; Gravesen, 1978). Dust mites are very sensitive to relative humidity (RH), with greatest survival between 70% and 85% RH (Arlian, 1975; Colloff, 2009). When humidity falls below ~50%, dust mites die because their only mechanism for water intake is via their exoskeleton, so they become severely dehydrated.

Temperate and tropical areas differ in the distribution of dust mite taxonomic groups (i.e., taxa), and there can also be differences in patterns of sensitisation to their allergens among allergic individuals living in those areas. For example, the most abundant mite species recovered in house dust from Puerto Rico in studies to date has been *Dermatophagoides pteronyssinus*, followed by *Blomia tropicalis* and *D. farinae*, among other taxa (Montealegre *et al.*, 1997a). These findings contrast with those from more temperate climates, where allergens from *D. farinae* are more frequently recovered (Chew *et al.*, 1998; Rose *et al.*, 1996; Wood *et al.*, 1988). In one study conducted in Puerto Rico, 61% of an asthmatic population was sensitised to at least one species of dust mite compared to 22% of the control population (Montealegre *et al.*, 1997b). In particular, more of the Puerto Rican individuals with asthma had serum immunoglobulin E (IgE) that was specific to allergens from *B. tropicalis* than from *D. pteronyssinus*, reflecting the possible higher allergenicity of the *B. tropicalis* allergens (Arruda *et al.*, 1997). In another Puerto Rican study, 46% and 47% of asthmatic children were sensitised to *D. farinae* and *D. pteronyssinus*, respectively (Nazario *et al.*, 2000). In the mainland United States, only 35% of asthmatic children in urban areas were found to have dust mite allergy (as measured by extracts of *D. pteronyssinus* or *D. farinae*; Rosenstreich *et al.*, 1997). The lower prevalence of dust mite allergy among individuals with asthma in colder US urban areas may be due in part to their lower exposures. High levels of dust mite allergen were found in only 10% of homes in the asthmatic children living in the urban US compared with 100% of homes in Brazil and 20% of homes in Colombia (Arruda *et al.*, 1991; Fernández-Caldas *et al.*, 1993; Rosenstreich *et al.*, 1997). An interesting paradox is that in a population-based study of the United States, dust mite allergen levels were higher in the colder northeast, followed by the warmer more humid south, then the midwest, and west (geometric mean = 3.4, 1.8, 1.5, and 0.3 μg/g, respectively; $p<0.001$; Arbes *et al.*, 2003). This finding suggests that merely looking at a geographic region's climate could mask other factors that lead to a dust mite population's ability to thrive in the indoor environment.

7.3 Cockroach and Mouse Allergens

Cockroaches and mice mainly have allergens in their faeces and urine, respectively, although mouse dander can also contain allergens (Phipatanakul *et al.*, 2012; Portnoy *et al.*, 2013). Thus far, only the allergens from German and American cockroaches have been well characterised, but there are several allergens which seem to cross-react not only with those of other cockroaches but with those of other insects (Portnoy *et al.*, 2013). The mouse urinary proteins are mainly characterised from the *Mus musuclus* species (the house mouse), but it is conceivable that the allergens cross-react with those of other mice (e.g., *Peromyscus* spp. and *Apodemus* spp.), given that there is about 80% homology with rat allergens (Phipatanakul *et al.*, 2012). Both cockroaches and mice can have domiciliary tendencies which can lead to infestations of buildings if given enough shelter, food, and water (Phipatanakul *et al.*, 2012; Portnoy *et al.*, 2013). Therefore, the concentrations of indoor allergens are likely not as susceptible to direct changes in climate. What is unknown is whether the geographic distribution of cockroach and mouse species will change as higher latitude climates become warmer.

Only in the past decade have researchers investigated the home environment for associations between cockroach and mouse allergens and development of childhood allergy and asthma (Chew *et al.*, 2008; Matsui *et al.*, 2003, 2004, 2005, 2006; Phipatanakul *et al.*, 2000a, 2000b, 2005). Furthermore, there is sparse literature on the association between these allergens and asthma among the elderly (Chew *et al.*, 2006a). Part of the reason for the dearth of older literature is that standardised immunoassays for cockroach and mouse allergens were developed after the dust mite and cat allergen assays. Another reason could be that cockroach and mouse allergies were considered mainly an urban issue, so the allergens were not routinely measured in many of the earlier large asthma studies (Martinez *et al.*, 1995; Peat *et al.*, 1996; Sears *et al.*, 1989).

7.4 Epidemiologic Studies of Dust Mite, Cockroach, and Mouse Allergic Sensitisation

Table 7.1 shows a summary of studies of sensitivity to dust mite, cockroach, and mouse allergens where at least two of the allergens were assessed (e.g., dust mite and cockroach, or cockroach and mouse). The reason for listing these studies is to enable comparisons among the three allergens (i.e., to gauge relative importance in terms of sensitisation) and to give examples of sensitisation prevalence now (1997–2012) and how these patterns might change in the future. Many of the studies focussed on children; this is not necessarily a limitation. In fact, it makes it easier to assume that the home exposures are spatially and temporally associated with children's allergies compared with those of adults who might have moved several times and perhaps to drastically different climatic regions and housing stock over their lifetime. As shown in Table 7.1, dust mite allergy is often dominant in prevalence (but not necessarily severity), yet the prevalence of cockroach allergy can approach and sometimes surpass that of dust mites in some communities. While not as common as dust mite and cockroach allergy, 28% of US inner city children had mouse allergy (Gruchalla, 2000). Specifically, 38% of children from the Bronx and 33% from other urban areas of New York had mouse allergen sensitivity

Table 7.1. *Dust mite, cockroach, and mouse allergic sensitisation.*

Authors*	Study Parameters	Allergic Sensitisation Prevalence (% positive)		
		Dust mite	Cockroach	Mouse
Rosenstreich *et al.*, 1997	Skin prick tests of 476 asthmatic children from eightww US inner cities	35	37	Not measured (see Phipatanakul *et al.*, 2000c)
Phipatanakul *et al.*, 2000c	Skin prick tests of 499 asthmatic children from eight US inner cities	Not measured	Not measured	18
Gruchalla, 2000	Skin prick tests of 942 asthmatic children (5–11 yr) from seven US inner cities	57	69	28
Matsui *et al.*, 2004	Skin prick tests of 335 asthmatic children (6–17 yr) from Baltimore	69	25	13
Arbes *et al.*, 2005	Skin prick tests of 10,508 individuals (6–59 yr) from a population-based study in the United States	28	26	Not measured
Chew *et al.*, 2008	IgE in serum from 341 preschool children (4 yr) in New York City	16	22	10
Mahesh *et al.*, 2010	Skin prick tests of 546 patients with allergic rhinitis or asthma (9–55 yr) in southern India	70% of urban, 68% of suburban, 65% of rural residents	53% of urban, 39% of suburban, 47% of rural residents	Not measured
Olmedo *et al.*, 2011	IgE in serum from 225 children (7–8 yr) in New York City	30% of asthmatic vs 15% of non-asthmatic children	24% of asthmatic vs 10% of non-asthmatic children	15% of asthmatic vs 6% of non-asthmatic children
Stevens *et al.*, 2011	IgE in serum from 181 children in Ghana	51% of asthmatic vs 16% of non-asthmatic children	59% of asthmatic vs 37% of non-asthmatic children	Not measured
Brunst *et al.*, 2012	Skin prick tests of 472 children (7 yr) from Cincinnati	12	4	Not measured
Perry *et al.*, 2012	Skin prick tests of 91 asthmatic children (4–17 yr) from rural Arkansas	27	14	15

* Studies are listed chronologically from oldest to newest.

(Gruchalla, 2000). Interestingly, sensitivity to mouse allergens in laboratory workers having contact with mice has been reported to range only between 7% and 32% (Schumacher *et al.*, 1981; Venables *et al.*, 1988). What is not shown in Table 7.1 but merits attention is that the combination of allergic sensitivity to cockroach and high levels (>8 U/g) of cockroach allergen (Bla g 1) in house dust was significantly associated with increased numbers of hospitalisations, missed school days, and days of wheezing among asthmatic children (Rosenstreich *et al.*, 1997).

7.5 Home Characteristics Associated with Dust Mite, Cockroach, and Mouse Allergens

It is widely accepted that exposure and sensitisation to allergens from dust mites, cockroaches, and mice are common in urban, suburban, and rural areas around the world (Custovic *et al.*, 2003; de Blay *et al.*, 1997; Illi *et al.*, 2006; Lane *et al.*, 2005; Mahesh *et al.*, 2010; Perry *et al.*, 2012; Perzanowski *et al.*, 2002; Platts-Mills *et al.*, 1997; Stelmach *et al.*, 2002; Tovey *et al.*, 2000; van Strien *et al.*, 1994; Wahn *et al.*, 1997; Wickens *et al.*, 2001). As mentioned earlier, the main factor required for proliferation of dust mites is sufficient humidity, and the main factors for cockroaches and mice are adequate sources of water and food. However, additional characteristics can affect the levels of these allergens in the home. Some of these factors include neighbourhood and individual home characteristics. For example, dust mites thrive in plush surfaces (e.g., carpet, beds; Platts-Mills *et al.*, 1989). Chew *et al.* (1999) found that within the same climatic region (i.e., northeast United States), dust mite allergen levels can vary greatly between houses in the suburbs and apartments in the city. They stated that the low dust mite levels in apartments were due to apartment dwellers having less control over their overheated apartments during winter which decimated the dust mite population (Chew *et al.*, 1999). This situation of increased indoor residential temperatures during winter is not unique to the United States. In the UK, from 1978 to 1996, indoor residential temperatures have increased an estimated 1.3°C per decade (Mavrogianni *et al.*, 2013). In a larger study in the same northeast US area, researchers observed that carpet remained a significant risk factor for high levels of dust mite allergen, even in overheated apartments (Chew *et al.*, 1998). Figure 7.1 explains why carpeting can be a risk factor.

Another example of the influence of housing stock and location is evident with cockroach and mouse allergens. Even within a single city, different neighbourhoods can have varying levels of these allergens (Chew *et al.*, 2003a; Olmedo *et al.*, 2011; Rosenfeld *et al.*, 2010). Possible reasons include density of restaurants, level of housing disrepair, and height and density of apartments within a building. For example, Chew *et al.* (2003a) reported that apartments in the Harlem neighbourhood of New York (where many high-rise housing complexes are separate from restaurants and grocery stores) did not have a high prevalence of mouse allergen. However, the apartments in the Washington Heights neighbourhood (where many high-rise housing buildings have retail space with restaurants and grocery stores on the ground floors) had a high prevalence of mouse allergen. For these reasons (and possibly others), it is difficult to speculate about how climate change can affect indoor allergens. However, if residents start migrating more toward high-density urban areas (e.g., as an adaptation to avoid high energy costs associated with transportation and/or heating/cooling single-family homes), then exposures to cockroach and mouse allergens might become more relevant and allergies to them could possibly equal or surpass those to dust mites as is already seen in several inner cities of the United States (Gruchalla *et al.*, 2005). This scenario, along with potential synergistic effects between allergens and air pollution (Chapter 8), could mean that allergen avoidance strategies might need to be retooled in order to be effective for future generations.

7.6 Fungi and Fungal Allergens

Fungal allergens can be located in spore walls and intracellular components or they can be released during germination and growth processes (Green *et al.*, 2003, 2005; Horner *et al.*, 1995; Mitakakis *et al.*, 2001). This means that dead, dormant, and living spores and fungal fragments can contain allergens. To date, several species of fungi have well-characterised allergens (Horner *et al.*, 1995; Simon-Nobbe *et al.*, 2008), yet only one large-scale environmental assessment study has measured fungal allergens (Salo *et al.*, 2005). Because there is much cross-reactivity among fungal allergens, it is difficult to associate specific allergens, such as those from *Alternaria alternata*, to adverse respiratory effects. The cross-reactivity is due in part to the varied biologic functions of fungal allergens which range from heat shock proteins to enzymes involved with tissue infection (Simon-Nobbe *et al.*, 2008).

A discussion of fungal allergens would not be complete without addressing other components of fungi that can give rise to adverse health effects. Fungal exposure in the indoor environment can comprise not only allergens but also concomitant exposure to microbial volatile organic compounds (VOCs), inflammatory agents such as (1-3)-β-D-glucans, and in some cases mycotoxins which inhibit protein synthesis. Thus far, only the glucan component has been evaluated extensively in the exposure and respiratory health studies (Douwes, 2005; Gehring *et al.*, 2001; Iossifova *et al.*, 2007), but there is growing evidence for a link between microbial VOCs (e.g., 3-methylfuran, mouldy odour) and respiratory symptoms (Jaakkola *et al.*, 2005; Wålinder *et al.*, 2005). Whether or not specific fungal allergens are

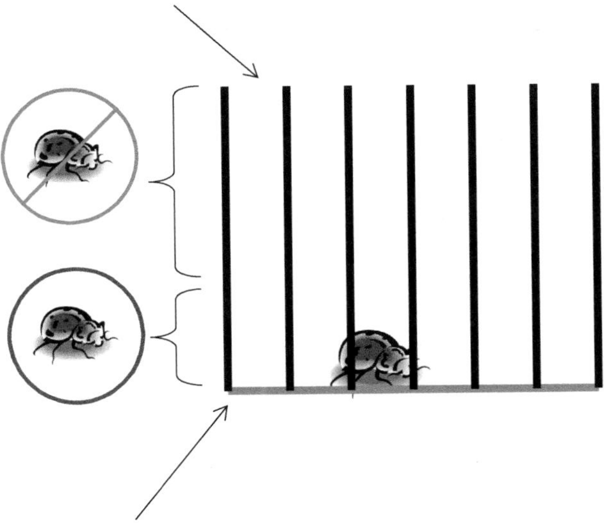

Figure 7.1. An explanation for why carpet serves as a reservoir for dust mites.

associated with respiratory symptoms independent of the other fungi-derived components remains unknown. However, there is consensus that fungal sensitisation is associated with asthma and other respiratory symptoms (Halonen *et al.*, 1997; O'Hollaren *et al.*, 1991; Perzanowski *et al.*, 1998; Pongracic *et al.*, 2010; Salo *et al.*, 2011).

Sensitisation to fungi has been identified in several multi-centre studies to be a risk factor for asthma (Zureik *et al.*, 2002), and the prevalence of hypersensitivity to common fungal allergens among atopic patients has been estimated to range anywhere from as low as 2% to as high as 90% (Black *et al.*, 2000; Brunekreef *et al.*, 1989; Hasnain *et al.*, 1985; Helbling *et al.*, 1994; Horner *et al.*, 1995; Lehrer *et al.*, 1986; O'Hollaren *et al.*, 1991; Portnoy *et al.*, 1987; Santilli Jr *et al.*, 1985, 1990; Sprenger *et al.*, 1988; Szánthó *et al.*, 1992; Vijay and Kurup, 2004; Zureik *et al.*, 2002). As with most allergens, variations in sensitisation are dependent upon differences in exposures between different sampling environments, the source and batch of commercial extracts, the selection criteria of test subjects, and the methods of analysis (Esch, 2004). Attempts to relate fungal exposure to development of fungal sensitisation are further complicated by the subjective methods used in quantifying predicting parameters, which include spore counts, surveys, and other indoor assessments of mould (common term for filamentous fungi) that are commonly used in statistical analyses to determine the risk factors for mould-induced respiratory allergic disease. As a result of these factors, the interpretation of personal exposure to fungal allergens has been restricted to the inhalation of fungal spores from a small and select number of fungi, including *Alternaria*, *Aspergillus*, *Penicillium*, and *Cladosporium* species for reasons that relate to their airborne abundance in many geographic locations and morphologically recognisable phenotypic features (Cruz *et al.*, 1997). This is particularly the case for *Alternaria*, which has been widely studied in epidemiologic investigations, due to the ease of identifying its large and visibly distinctive spores that have been shown in many environments to exacerbate rhinitis (Andersson *et al.*, 2003) and asthma (Downs *et al.*, 2001; Fung *et al.*, 2000; Halonen *et al.*, 1997; Licorish *et al.*, 1985; Negrini *et al.*, 2000; Neukirch *et al.*, 1999; O'Hollaren *et al.*, 1991).

7.7 Direct Effects on Indoor Allergens

7.7.1 Hurricanes and Floods

As earlier described in the most recent IPCC assessment report (IPCC, 2013), more intense precipitation is likely to occur in areas of the world which are currently considered temperate. Whether or not the precipitation arrives as snow or rain, the indoor environment could be impacted directly. The obvious direct impact is that intense rain can lead to floods and then the indoor environment is contaminated with microbial-laden flood water. Another important but perhaps more indirect effect is that fungal spores indoors that were heretofore kept from growing are now provided with enough moisture to enable widespread fungal proliferation throughout the indoor environment. A less obvious example is when snow melts on a warm roof during the day and then refreezes during the evening and forms ice dams. The next time the ice or snow begins to melt, ice dams can cause water to flow backward underneath shingles and into the building. The trapped water provides enough moisture to promote fungal growth. Although there is not much published literature on the effect of ice dams and fungal-related health effects, there is a growing body of literature on effects of floods and hurricanes (Grimsley *et al.*, 2012; Rabito *et al.*, 2008, 2010; Ross *et al.*, 2000, 2002). This dearth of publications notwithstanding, ice dams probably lead to a more long-term hidden mould problem which could make it difficult to examine temporality of exposures and health effects.

Not all flooding is the same for fungi. When flood waters quickly recede, buildings can be dried faster. However, flood waters that do not recede quickly or homes

Figure 7.2. New Orleans building after flooding from Hurricane Katrina (photo by Ginger Chew).

that have basements that hold water as if they were indoor swimming pools can experience a different profile and magnitude of mould growth. A common scene in such homes is a water line observed on the drywall (plasterboard), below which little mould growth is observed (Figure 7.2). There are two possible things that explain this lack of mould growth below the water line. First, almost all moulds need oxygen and cannot sporulate in liquid. The lower half of the drywall had less mould growth because it was underwater for a sustained period of time. Second, flood waters contain the chemicals found in the home itself (e.g., bleach, pesticides, and other cleaning products) plus the chemicals from the soil outside and possibly other toxicants that occur from local industrial or agricultural sources. Many of the chemicals can be fungal inhibitors or can be fungicidal, which could limit mould growth in areas below the water line. Perhaps a combination of these reasons can explain this mould growth pattern in buildings that have endured long-term flooding.

Few major flood events have been examined for their impact on fungal exposure and respiratory health effects (Grimsley *et al.*, 2012; Rabito *et al.*, 2008, 2010; Ross *et al.*, 2000, 2002). The Mississippi River flood study examined asthma morbidity among a wide age range of residents (5–49 years; Ross *et al.*, 2000). However, the Hurricane Katrina studies focused on allergy and asthma among children. For both major flood events, the studies observed little association between mould exposure and respiratory symptoms. A main reason for this lack of association is the variability in both exposures and symptoms. Furthermore, the populations that remain in the vicinity after a major flood event might not be representative of all affected residents. For example, those who have the resources to fix their homes quickly (i.e., the healthy rebuilding effect) might have less mould exposure and therefore fewer symptoms (Barbeau *et al.*, 2010). On the other hand, poorer residents who remain but do not receive assistance to quickly fix their homes might experience higher and more long-term mould exposure.

After the 1993 Mississippi River flood, Ross *et al.* (2000) assessed mould spores, lung function, and respiratory symptoms for fifty-seven asthmatic residents living in forty-four homes in Illinois in April through October 1994. The mould spore average was 2,190 spores m^{-3}. The researchers found that high *Alternaria* concentrations were associated with almost five-fold higher odds of missing sleep due to asthma. In their second analysis of the data from the Mississippi River floods, the researchers had a slightly different sample size; they assessed mould spores, lung function, and respiratory symptoms for a subset of forty asthmatic residents (who remained in the study for follow-up) in spring through fall 1994 (Ross *et al.*, 2002). The mould spore average was 5,692 spores m^{-3}. The researchers found that higher mould spore concentrations were associated with improved peak expiratory flow rate (PEFR) and respiratory symptom scores. The authors attribute part of the paradoxical results to potential bias in self-reported diary cards for PEFR and symptoms.

The inverse relationship between indoor air mould concentrations and asthma symptoms was also observed in a study conducted in New Orleans 18 months after Hurricane Katrina (Grimsley *et al.*, 2012). Children aged 4–12 years with moderate to severe asthma (n = 182) were tested for inhalant allergies (e.g., dust mite, cat, dog, cockroach, mouse, and the mould *Alternaria alternata*) and their homes were sampled for indoor allergens, endotoxin, and mould at three time points throughout the year (baseline, 6 months, and 12 months). The mould spore average was 501 spores m^{-3}. Although there were no significant associations between indoor air mould concentrations and skin prick test sensitivity, the authors did observe an inverse relationship between indoor air mould concentrations and asthma symptom days during the baseline home visit. This relationship was not consistently inverse nor was it significant for the 6- and 12-month follow-up visits, so the study authors were dismissive of the inverse relationship. However, this finding at baseline taken along with the second analysis by the Mississippi River flood researchers warrants further investigation. It could be that there are neighbourhood-level effects (e.g., healthy rebuilding effect) which temporarily increase outdoor mould levels (and thus indoor mould levels which are often affected by outdoor levels; Chew *et al.*, 2003b) such that children living in homes located within these neighbourhoods appear healthier than those in new homes (with low indoor mould levels) that are located in neighbourhoods without much rebuilding.

Rabito *et al.* (2008) conducted two separate studies of mould exposure in post–Hurricane Katrina New Orleans. In the first, the study site was a school which had the ability to reopen in January 2006 (only 5 months after the hurricane). Respiratory health questionnaire and lung function data were collected on children aged 7 to 14 years, and air sampling for fungi in their homes was conducted at baseline and again after 2 months. The 75th percentile for mould concentration was 100 colony-forming units (cfu) m^{-3} and 70 cfu m^{-3} at the two time points. The levels were several orders of magnitude lower than those reported in unoccupied homes immediately after the hurricane (Chew *et al.*, 2006b). Nonetheless, there was an overall decrease in mould levels and respiratory symptoms over the study period, and indoor mould levels were low despite reported hurricane damage. Although many of the homes had sustained hurricane damage, the authors stressed that their results might not be generalisable to the residents of other homes who did not have the financial means to return to the city and to either repair their homes or relocate to a non-flooded area.

In another study by Rabito *et al.* (2010), patients in an allergy clinic were enrolled from winter 2005 to winter 2008 (Rabito *et al.*, 2010). For 529 patients, mould exposure was assessed by questionnaire and mould allergy was assessed by skin prick test. Mould exposure (in terms of extent of damage or duration of exposure) was not associated with mould allergy. This finding was similar to the Grimsley *et al.* (2012) study mentioned above which also found that mould exposure was not associated with an increase in mould allergy. The authors acknowledged that minorities and those without health insurance were underrepresented in their study, thus limiting generalisability.

7.7.2 Warmer Weather

In regions which become warmer and with increasing absolute humidity (e.g., from increased precipitation events), the profile of dust mites could change such that more tropical dust mites could proliferate in areas which are not currently tropical. This is due to a combination of people and furnishings travelling (which can passively import the 'seed populations' of tropical mites) and subsequently the ability of the relocated mites/mite eggs to flourish in a warm humid environment. On the other hand, it is more difficult to project what will happen in regions that experience an increase in temperature without a concomitant increase in absolute humidity. For example, it is clear that arid regions such as New Mexico are

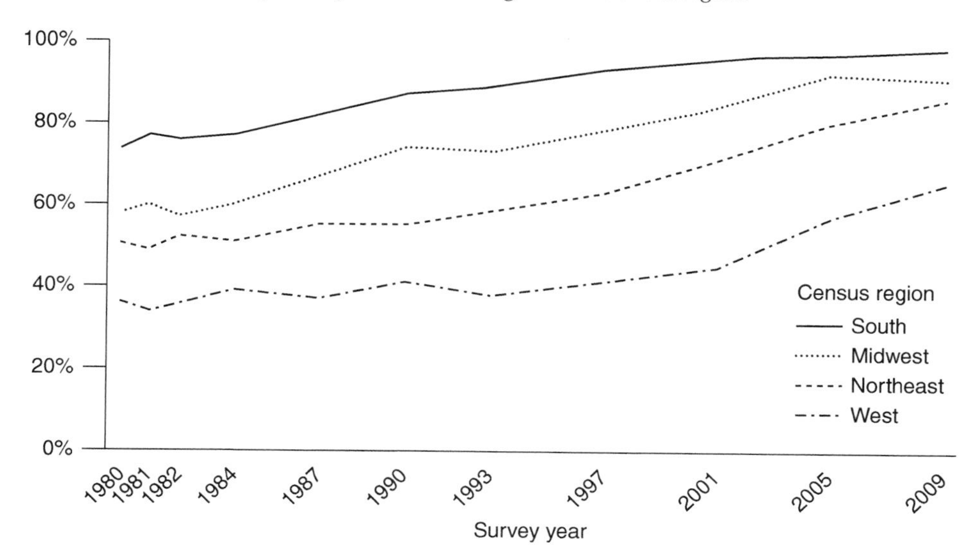

Figure 7.3. Trends in the percentage of homes with air-conditioning across the United States from 1980 to 2009.
Source: Energy Information Administration (2011).

expected to get hotter, but even though warm air can hold more water vapour, the RH can be rather low and will not sustain a dust mite population (Ingram *et al.*, 1995). The more difficult example is a temperate region such as New York City. As New York's temperature increases, the air in this coastal city might be able to hold more water vapour and thus facilitate tropical mites moving to this urban environment; however, there are adaptation measures of the residents (namely air-conditioning) that might preclude this scenario.

7.8 Indirect Effects on Indoor Allergens

7.8.1 More Air-Conditioning and/or Dehumidifier Use in Summer

In the United States, air-conditioning use has steadily increased (Figure 7.3); therefore, it is likely that this adaptation measure will impact dust mite populations.

The reason for a possible decrease in dust mite populations is best seen in past studies from the United States and Europe. In a study that covered a wide geographic area of the United States, air-conditioning was independently associated with lower dust mite allergen levels (Lintner and Brame, 1993). In a state with a temperate climate (Ohio), researchers found that among three groups of homes (those without air-conditioning, those with air-conditioning only, and those with air-conditioning and a high-efficiency dehumidifier) the dust mite allergen levels

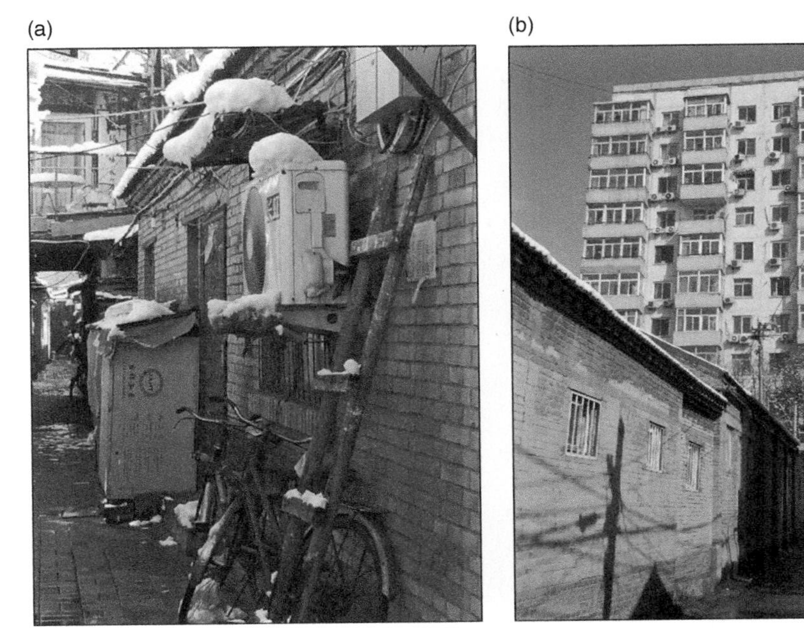

Figure 7.4. Heat pumps for air-conditioning in summer and heating during winter are widely used in Beijing, China. They are shown in these pictures attached to the sides of the buildings (photos by Ginger Chew).

in the homes with both air-conditioning and dehumidifiers were more than ten times lower that the other homes (Arlian *et al.*, 2001). Given that the outdoor temperature in the Ohio summer months was ~25°C, it is not clear how frequently the air-conditioning was used in the air-conditioning only group. For this reason alone, the dehumidification (with air-conditioning only) might not have been as much as could be expected in a scenario where Ohio summers would be longer and warmer. Furthermore, Swedish researchers found that with energy-efficient housing, mechanical exhaust and adequate supply ventilation can significantly decrease mite allergen (Sundell *et al.*, 1995). The implication of widespread air-conditioning use in other countries such as China (Figure 7.4) would be a decrease in indoor dust mite allergen exposure, and this could potentially decrease the ranking of dust mites as a major indoor allergen. Whether or not this would also change the allergic sensitisation patterns remains unknown. Conceivably, fewer people would be allergic to dust mites (similar to those in the Italian Alps or in a US city with high elevation and low humidity, such as Denver), so patients with allergic asthma would mount immune responses to other things in their environment (e.g., cockroaches, cats, mice; Boner *et al.*, 2002; Gruchalla *et al.*, 2005).

Figure 7.5. Flooded basements are difficult to dry quickly. Since they are located below grade (i.e., below ground), the water remains pooled inside. Also, many basements do not have windows, so even after all water-damaged furnishings and building materials are removed, ventilation is not as effective as it is for rooms with windows (photo by Carl Grimes from Healthy Habitats, LLC).

7.8.2 More Humidifier Use in Summer

Humidifiers are an adaptation strategy for those in an arid environment, but it is unlikely that residents could use the humidifiers frequently enough to keep the humidity high enough to sustain the dust mite population; however, there is some evidence that this is possible (Prasad *et al.*, 2009). Prasad *et al.* (2009) showed that evaporative coolers can increase indoor RH to a point where dust mite sensitisation increases. The researchers posit that it is an increase in dust mite allergen in the environment that is tied with the increase in sensitisation, but no allergen measurements were collected. Nonetheless, it remains an interesting question that warrants further investigation.

7.9 Vulnerable Populations

During major flooding events, homes with basements often take the longest time to dry because the water pools inside, producing suitable conditions for mould growth (Figure 7.5). Thus, residents in these homes can be considered a vulnerable

Table 7.2. *Percentage of single-unit houses with basement (full or partial) built within the four years prior to the American Housing Survey for each of the six metropolitan statistical areas.*

MSA	AHS Survey Year	% of New single-unit Houses with Basement	AHS Survey Year	% of New Single-unit Houses with Basement	Name and Year of Hurricane
Atlanta	1998	52	2011	33	Ivan, 2004
Baltimore	1998	79	2007	87	Floyd, 1999 and Charley, 2004
Memphis	1996	2	2011	0	Ivan, 2004
Boston	1998	92	2007	100	n/a
New York	1999	89	2009	72	n/a
Northern New Jersey	1999	61	2009	60	n/a

This information was obtained from the AHS metropolitan area databases (available at www.census.gov/housing/ahs/data/metrotext.html).

population to adverse health impacts from flooding. In a very preliminary exercise, we (the authors of this chapter) attempted to assess the extent of this vulnerable population. We obtained county-level estimates of housing structures in 100-year flood hazard areas for the entire United States as measured by the Federal Emergency Measurement Agency (FEMA). We classified the counties into quintiles based on the national distribution of housing units in the 100-year flood hazard areas. We then produced the choropleth map to show those county classifications (Figure 7.6). Having examined the flood hazard areas by counties, we wanted to juxtapose the distribution of housing units with basements with that information. We identified six metropolitan statistical areas (MSAs) from the population-based American Housing Survey (AHS) to determine houses with basements for two consecutive survey years (roughly a decade apart). The housing surveys contained information on (i) single-unit houses with basements and (ii) single-unit houses built within the last 4 years when the survey was conducted that had basements.

Within three of these MSAs, percentage of new single-unit homes with basements before and after major hurricane/flood events were examined to assess whether there appeared to be a shift in building fewer homes with basements (Table 7.2). Three MSAs in the northeast (where no major hurricanes occurred during the time period of interest) were included for comparison. The reason for this analysis was two-fold: (1) to derive preliminary evidence on whether the proportion of new homes with basements being built after hurricane was changing and (2) to show the current vulnerability of MSAs that could sustain damage in

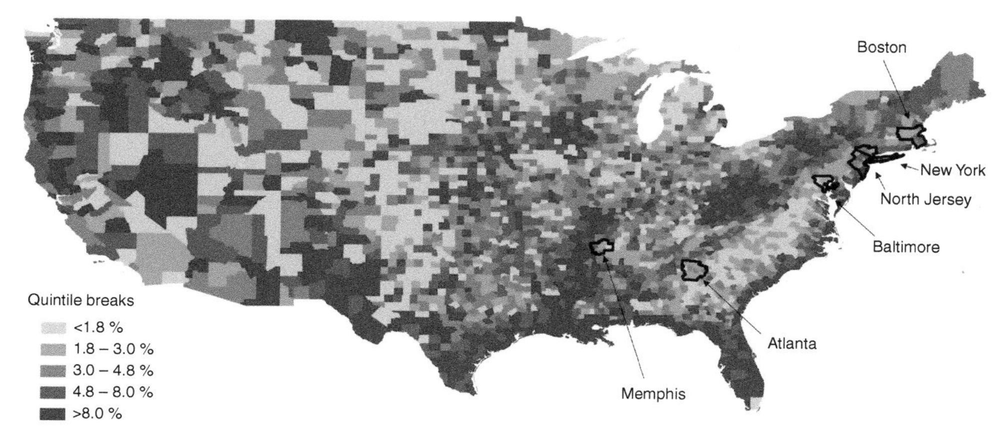

Figure 7.6. US map representing percentage of housing units in each county in 100-year flood hazard areas (data obtained from FEMA: hazards.fema.gov/femaportal/wps/portal/NFHLWMS). The colour codes represent quintiles calculated based on the distribution of percentage of housing units in the flood hazard area. Counties shaded with darker grey indicate a higher proportion of housing units that are in flood hazard areas compared to those in lighter grey which indicate a relatively lower percentage of housing units in flood hazard areas. The map also shows the six MSAs selected based on availability of housing structure information from the AHS. (A black and white version of this figure will appear in some formats. For the colour version, please refer to the plate section.)

future flood events. Table 7.2 suggests that while some MSAs showed a decrease in proportion of new buildings with basements (e.g., Atlanta, New York), there were other MSAs where it stayed the same (e.g., northern New Jersey) or increased (e.g., Baltimore, Boston). Basements which are repeatedly flooded can pose a risk not only in terms of increased exposure to indoor allergens but also to the chemicals used in the remediation process.

Using the most current round of the AHS data for each MSA, we estimated the proportion of single-unit houses with basements. We then mapped the percentage of housing units estimated to be in 100-year flood hazards and compared the proportion of single-unit houses with basements to assess vulnerability (Figures 7.6 and 7.7). The maps in Figure 7.7 show that even within metropolitan areas, disparities in vulnerability can exist. For example, 25.6% of homes in Ocean County, northern New Jersey, are in the 100-year flood zone compared with only 2.8% of homes in Middlesex County, northern New Jersey. This ten-fold difference might be offset by homes in Ocean County being built without basements. In fact, many homes in Ocean County do not have basements; they are built on elevated pylons. However, the AHS does not have the granularity to estimate county-level percentages of homes with basements; it can only estimate this at the MSA level (i.e., 72%). If Ocean County's percentage of homes with basements is similar to

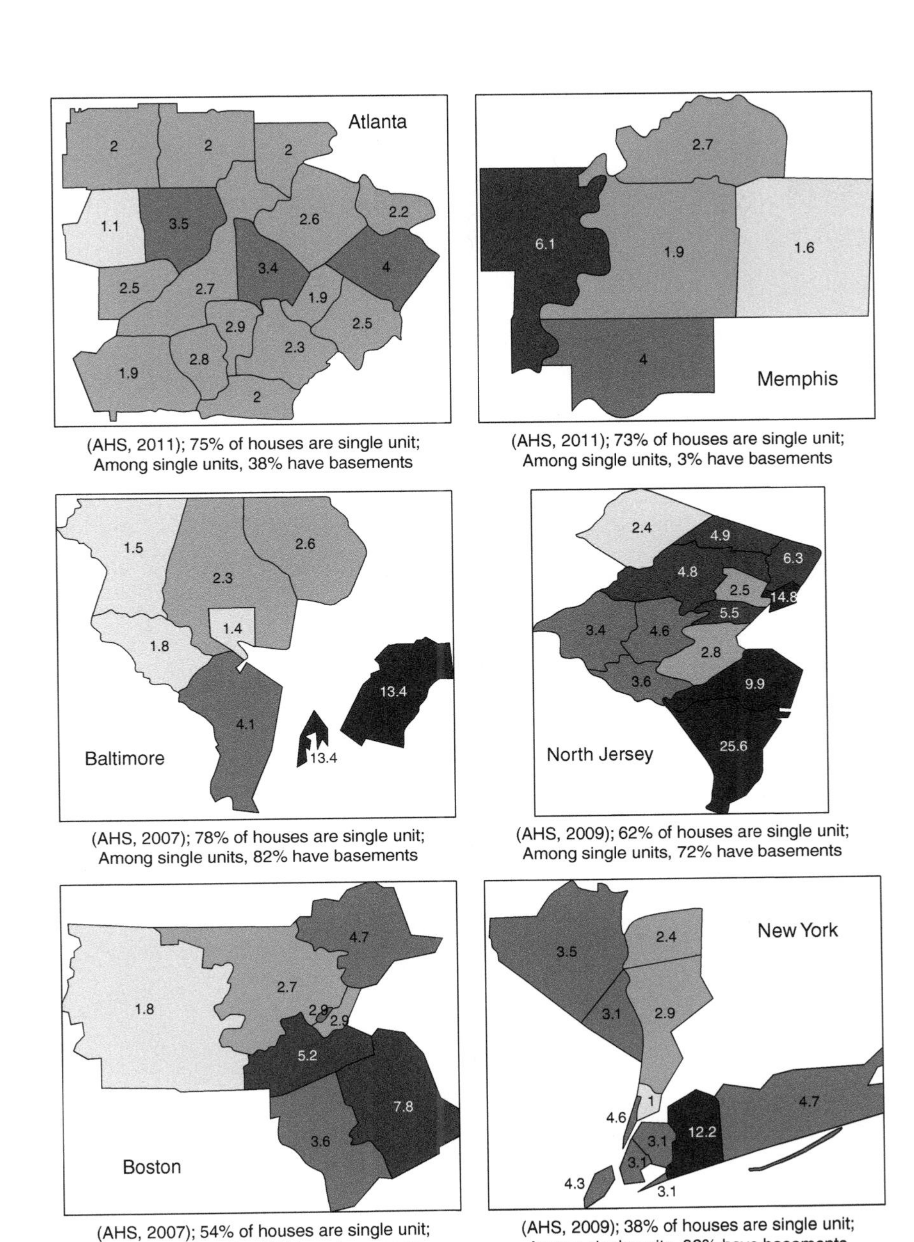

Figure 7.7. County-level estimates of percentage of housing units in 100-year flood hazard areas for six metropolitan statistical areas – Atlanta, Baltimore, Boston, Memphis, northern New Jersey, and New York. The colour codes represent quintiles calculated based on percentage of housing units in a 100-year flood hazard area for all counties in the United States. The counties included in each of the MSAs are the same as those used by the AHS. The estimates of percentage of single-unit houses among total housing units, and percentage of single-unit houses with basements (full or partial) are derived from the AHS metropolitan area database (available at www.census.gov/housing/ahs/data/metrotext.html).

that of the Middlesex County, then this could be a focus for public health education and perhaps also building code modifications to mitigate adverse effects of climate change.

7.10 Conclusions

At this time, many of the effects of climate change on indoor allergens are difficult to tease apart from the effects of global increases in building-related factors that might occur even without climate change (e.g., more residents using air-conditioning). However, there are some clear examples from past and recent literature to indicate that indoor environments (often influenced by outdoor conditions of a given region, locality, or neighbourhood) can affect dust mite, cockroach, and mouse allergens. This can occur, for example, by residents responding to overheated apartments in winter by opening their windows and subsequently drying out the air to a point where dust mites cannot live. Finally, the projected increase in precipitation of many areas will inevitably lead to more flooding. This is perhaps the clearest example of how climate change can affect the indoor environment with respect to fungal growth and their associated allergens and other inflammatory components. There are currently several surveys which can track how housing factors change over time (e.g., AHS, Residential Energy Consumption Survey) and there are model projections for flood areas. Some combination of these efforts will shed light on how climate change could affect indoor allergens in the future.

References

Andersson, M., Downs, S., Mitakakis, T., Leuppi, J., Marks, G. (2003). Natural exposure to *Alternaria* spores induces allergic rhinitis symptoms in sensitized children. *Pediatric Allergy and Immunology*, 14(2), 100–105.

Arbes Jr, S. J., Cohn, R. D., Yin, M., *et al.* (2003). House dust mite allergen in US beds: results from the first National Survey of Lead and Allergens in Housing. *The Journal of Allergy and Clinical Immunology*, 111(2), 408–414.

Arbes Jr, S. J., Gergen, P. J., Elliott, L., Zeldin, D. C. (2005). Prevalences of positive skin test responses to 10 common allergens in the US population: results from the Third National Health and Nutrition Examination Survey. *The Journal of Allergy and Clinical Immunology*, 116(2), 377–383.

Arlian, L. G. (1975). Dehydration and survival of the European house dust mite, *Dermatophagoides pteronyssinus*. *Journal of Medical Entomology*, 12(4), 437–442.

Arlian, L. G., Neal, J. S., Morgan, M. S., *et al.* (2001). Reducing relative humidity is a practical way to control dust mites and their allergens in homes in temperate climates. *The Journal of Allergy and Clinical Immunology*, 107(1), 99–104.

Arruda, L. K., Rizzo, M. C., Chapman, M. D., *et al.* (1991). Exposure and sensitization to dust mite allergens among asthmatic children in São Paulo, Brazil. *Clinical and Experimental Allergy*, 21(4), 433–439.

Arruda, L. K., Vailes, L. D., Platts-Mills, T. A. E., *et al.* (1997). Sensitization to *Blomia tropicalis* in patients with asthma and identification of allergen Blo t 5. *American Journal of Respiratory and Critical Care Medicine*, 155(1), 343–350.

Asher, M. I., Keil, U., Anderson, H. R., *et al.* (1995). International study of asthma and allergies in childhood (ISAAC): rationale and methods. *European Respiratory Journal*, 8(3), 483–491.

Barbeau, D. N., Grimsley, L. F., White, L. E., El-Dahr, J. M., Lichtveld, M. (2010). Mold exposure and health effects following hurricanes Katrina and Rita. *Annual Review of Public Health*, 31, 165–178.

Black, P. N., Udy, A. A., Brodie, S. M. (2000). Sensitivity to fungal allergens is a risk factor for life-threatening asthma. *Allergy*, 55(5), 501–504.

Boner, A., Pescollderungg, L., Silverman, M. (2002). The role of house dust mite elimination in the management of childhood asthma: an unresolved issue. *Allergy*, 57(Suppl 74), 23–31.

Brunekreef, B., Dockery, D. W., Speizer, F. E., *et al.* (1989). Home dampness and respiratory morbidity in children. *American Review of Respiratory Disease*, 140(5), 1363–1367.

Brunst, K. J., Ryan, P. H., Lockey, J. E., *et al.* (2012). Unraveling the relationship between aeroallergen sensitization, gender, second-hand smoke exposure, and impaired lung function. *Pediatric Allergy and Immunology*, 23(5), 479–487.

Chew, G. L., Burge, H. A., Dockery, D. W., *et al.* (1998). Limitations of a home characteristics questionnaire as a predictor of indoor allergen levels. *American Journal of Respiratory and Critical Care Medicine*, 157(5), 1536–1541.

Chew, G. L., Carlton, E. J., Kass, D., *et al.* (2006a). Determinants of cockroach and mouse exposure and associations with asthma in families and elderly individuals living in New York City public housing. *Annals of Allergy, Asthma & Immunology*, 97(4), 502–513.

Chew, G. L., Higgins, K. M., Gold, D. R., Muilenberg, M. L., Burge, H. A. (1999). Monthly measurements of indoor allergens and the influence of housing type in a northeastern US city. *Allergy*, 54(10), 1058–1066.

Chew, G. L., Perzanowski, M. S., Canfield, S. M., *et al.* (2008). Cockroach allergen levels and associations with cockroach-specific IgE. *The Journal of Allergy and Clinical Immunology*, 121(1), 240–245.

Chew, G. L., Perzanowski, M. S., Miller, R. L., *et al.* (2003a). Distribution and determinants of mouse allergen exposure in low-income New York City apartments. *Environmental Health Perspectives*, 111(10), 1348–1351.

Chew, G. L., Rogers, C., Burge, H. A., Muilenberg, M. L., Gold, D. R. (2003b). Dustborne and airborne fungal propagules represent a different spectrum of fungi with differing relations to home characteristics. *Allergy*, 58(1), 13–20.

Chew, G. L., Wilson, J., Rabito, F. A., *et al.* (2006b). Mold and endotoxin levels in the aftermath of Hurricane Katrina: a pilot project of homes in New Orleans undergoing renovation. *Environmental Health Perspectives*, 114(12), 1883–1889.

Colloff, M. J. (2009). *Dust Mites*. Collingwood: CSIRO Publishing.

Cruz, A., Saenz de Santamaría, M., Martínez, J., *et al.* (1997). Fungal allergens from important allergenic fungi imperfecti. *Allergologia et Immunopathologia*, 25(3), 153–158.

Custovic, A., Simpson, B. M., Simpson, A., *et al.* (2003). Current mite, cat, and dog allergen exposure, pet ownership, and sensitization to inhalant allergens in adults. *The Journal of Allergy and Clinical Immunology*, 111(2), 402–407.

de Blay, F., Sanchez, J., Hedelin, G., *et al.* (1997). Dust and airborne exposure to allergens derived from cockroach (*Blattella germanica*) in low-cost public housing in Strasbourg (France). *The Journal of Allergy and Clinical Immunology*, 99(1), 107–112.

Douwes, J. (2005). (1→3)-β-*D*-glucans and respiratory health: a review of the scientific evidence. *Indoor Air*, 15(3), 160–169.

Downs, S. H., Mitakakis, T. Z., Marks, G. B., *et al.* (2001). Clinical importance of *Alternaria* exposure in children. *American Journal of Respiratory and Critical Care Medicine*, 164(3), 455–459.

Energy Information Administration (2011). Residential Energy Consumption Survey (RECS), 2009. Available at: www.eia.gov/consumption/residential/reports/2009/air-conditioning.cfm. Accessed 29 April 2016.

Esch, R. E. (2004). Manufacturing and standardizing fungal allergen products. *The Journal of Allergy and Clinical Immunology*, 113(2), 210–215.

Fernández-Caldas, E., Puerta, L., Mercado, D., Lockey, R. F., Caraballo, L. R. (1993). Mite fauna, *Der p* I, *Der f* I and *Blomia tropicalis* allergen levels in a tropical environment. *Clinical and Experimental Allergy*, 23(4), 292–297.

Fung, F., Tappen, D., Wood, G. (2000). *Alternaria*-associated asthma. *Applied Occupational and Environmental Hygiene*, 15(12), 924–927.

Gehring, U., Douwes, J., Doekes, G., *et al.* (2001). β(1→3)-glucan in house dust of German homes: housing characteristics, occupant behavior, and relations with endotoxins, allergens, and molds. *Environmental Health Perspectives*, 109(2), 139–144.

Gold, D. R., Acevedo-Garcia, D. (2005). Immigration to the United States and acculturation as risk factors for asthma and allergy. *The Journal of Allergy and Clinical Immunology*, 116(1), 38–41.

Gravesen, S. (1978). Identification and prevalence of culturable mesophilic microfungi in house dust from 100 Danish homes: comparison between airborne and dust-bound fungi. *Allergy*, 33(5), 268–272.

Green, B. J., Mitakakis, T. Z., Tovey, E. R. (2003). Allergen detection from 11 fungal species before and after germination. *The Journal of Allergy and Clinical Immunology*, 111(2), 285–289.

Green, B. J., Sercombe, J. K., Tovey, E. R. (2005). Fungal fragments and undocumented conidia function as new aeroallergen sources. *The Journal of Allergy and Clinical Immunology*, 115(5), 1043–1048.

Grimsley, L. F., Wildfire, J., Lichtveld, M., *et al.* (2012). Few associations found between mold and other allergen concentrations in the home versus skin sensitivity from children with asthma after Hurricane Katrina in the Head-off Environmental Asthma in Louisiana study. *International Journal of Pediatrics*, 2012, 427358.

Gruchalla, R. S. (2000). Allergy skin test results of 942 urban asthmatic children: the Inner City Asthma Study (ICAS). *The Journal of Allergy and Clinical Immunology*, 105(1), S368–S369.

Gruchalla, R. S., Pongracic, J., Plaut, M., *et al.* (2005). Inner City Asthma Study: relationships among sensitivity, allergen exposure, and asthma morbidity. *The Journal of Allergy and Clinical Immunology*, 115(3), 478–485.

Halonen, M., Stern, D. A., Wright, A. L., Taussig, L. M., Martinez, F. D. (1997). *Alternaria* as a major allergen for asthma in children raised in a desert environment. *American Journal of Respiratory and Critical Care Medicine*, 155(4), 1356–1361.

Hasnain, S. M., Wilson, J. D., Newhook, F. J. (1985). Fungi and disease: fungal allergy and respiratory disease. *New Zealand Medical Journal*, 98(778), 342–346.

Helbling, A., Reese, G., Horner, W. E., Lehrer, S. B. (1994). Aktuelles zur pilzsporenallergie [Current knowledge on fungal spore allergy]. *Schweizerische Medizinische Wochenschrift*, 124(21), 885–892.

Horner, W. E., Helbling, A., Salvaggio, J. E., Lehrer, S. B. (1995). Fungal allergens. *Clinical Microbiology Reviews*, 8(2), 161–179.

Illi, S., von Mutius, E., Lau, S., *et al.* (2006). Perennial allergen sensitisation early in life and chronic asthma in children: a birth cohort study. *The Lancet*, 368(9537), 763–770.

Ingram, J. M., Sporik, R., Rose, G., *et al.* (1995). Quantitative assessment of exposure to dog (Can f 1) and cat (Fel d 1) allergens: relation to sensitization and asthma among children living in Los Alamos, New Mexico. *The Journal of Allergy and Clinical Immunology*, 96(4), 449–456.

Iossifova, Y. Y., Reponen, T., Bernstein, D. I., *et al.* (2007). House dust (1–3)-β-D-glucan and wheezing in infants. *Allergy*, 62(5), 504–513.

IPCC (2013). *Climate Change 2013: The Physical Science Basis. Contribution of Working Group I to the Fifth Assessment Report of the Intergovernmental Panel on Climate Change* [Stocker, T. F., Qin, D., Plattner, G.-K., *et al.*, eds.]. Cambridge, UK and New York, NY: Cambridge University Press.

Jaakkola, J. J. K., Hwang, B.-F., Jaakkola, N. (2005). Home dampness and molds, parental atopy, and asthma in childhood: a six-year population-based cohort study. *Environmental Health Perspectives*, 113(3), 357–361.

Kitch, B. T., Chew, G., Burge, H. A., *et al.* (2000). Socioeconomic predictors of high allergen levels in homes in the greater Boston area. *Environmental Health Perspectives*, 108(4), 301–307.

Lane, J., Siebers, R., Pene, G., Howden-Chapman, P., Crane, J. (2005). Tokelau: a unique low allergen environment at sea level. *Clinical and Experimental Allergy*, 35(4), 479–482.

Lehrer, S. B., Lopez, M., Butcher, B. T., *et al.* (1986). Basidiomycete mycelia and spore-allergen extracts: skin test reactivity in adults with symptoms of respiratory allergy. *The Journal of Allergy and Clinical Immunology*, 78(3), 478–485.

Licorish, K., Novey, H. S., Kozak, P., Fairshter, R. D., Wilson, A. F. (1985). Role of *Alternaria* and *Penicillium* spores in the pathogenesis of asthma. *The Journal of Allergy and Clinical Immunology*, 76(6), 819–825.

Lin, S., Jones, R., Munsie, J. P., *et al.* (2012). Childhood asthma and indoor allergen exposure and sensitization in Buffalo, New York. *International Journal of Hygiene and Environmental Health*, 215(3), 297–305.

Lintner, T. J., Brame, K. A. (1993). The effects of season, climate, and air-conditioning on the prevalence of *Dermatophagoides* mite allergens in household dust. *The Journal of Allergy and Clinical Immunology*, 91(4), 862–867.

Mahesh, P. A., Kummeling, I., Amrutha, D. H., Vedanthan, P. K. (2010). Effect of area of residence on patterns of aeroallergen sensitization in atopic patients. *American Journal of Rhinology & Allergy*, 24(5), e98–e103.

Martinez, F. D., Wright, A. L., Taussig, L. M., *et al.* (1995). Asthma and wheezing in the first six years of life. *The New England Journal of Medicine*, 332(3), 133–138.

Matsui, E. C., Eggleston, P. A., Buckley, T. J., *et al.* (2006). Household mouse allergen exposure and asthma morbidity in inner-city preschool children. *Annals of Allergy, Asthma & Immunology*, 97(4), 514–520.

Matsui, E. C., Simons, E., Rand, C., *et al.* (2005). Airborne mouse allergen in the homes of inner-city children with asthma. *The Journal of Allergy and Clinical Immunology*, 115(2), 358–363.

Matsui, E. C., Wood, R. A., Rand, C., *et al.* (2003). Cockroach allergen exposure and sensitization in suburban middle-class children with asthma. *The Journal of Allergy and Clinical Immunology*, 112(1), 87–92.

Matsui, E. C., Wood, R. A., Rand, C., *et al.* (2004). Mouse allergen exposure and mouse skin test sensitivity in suburban, middle-class children with asthma. *The Journal of Allergy and Clinical Immunology*, 113(5), 910–915.

Mavrogianni, A., Johnson, F., Ucci, M., *et al.* (2013). Historic variations in winter indoor domestic temperatures and potential implications for body weight gain. *Indoor and Built Environment*, 22(2), 360–375.

Mitakakis, T. Z., Barnes, C., Tovey, E. R. (2001). Spore germination increases allergen release from *Alternaria*. *The Journal of Allergy and Clinical Immunology*, 107(2), 388–390.

Montealegre, F., Quiñones, C., Michelen, V., *et al.* (1997b). Prevalence of skin reactions to aeroallergens in asthmatics of Puerto Rico. *The Puerto Rico Health Sciences Journal*, 16(4), 359–367.

Montealegre, F., Sepulveda, A., Bayona, M., Quinones, C., Fernandez-Caldas, E. (1997a). Identification of the domestic mite fauna of Puerto Rico. *The Puerto Rico Health Sciences Journal*, 16(2), 109–116.

Nazario, S., Casal, J., Rodriguez, W., *et al.* (2000). Aeroallergen sensitivities in children attending public schools in San Juan, Puerto Rico. *The Journal of Allergy and Clinical Immunology*, 105(1), S235.

Negrini, A. C., Berra, D., Campi, P., *et al.* (2000). Clinical study on *Alternaria* spores sensitization. *Allergologia et Immunopathologia*, 28(2), 71–73.

Neukirch, C., Henry, C., Leynaert, B., *et al.* (1999). Is sensitization to *Alternaria alternata* a risk factor for severe asthma? A population-based study. *The Journal of Allergy and Clinical Immunology*, 103(4), 709–711.

O'Hollaren, M. T., Yunginger, J. W., Offord, K. P., *et al.* (1991). Exposure to an aeroallergen as a possible precipitating factor in respiratory arrest in young patients with asthma. *The New England Journal of Medicine*, 324(6), 359–363.

Olmedo, O., Goldstein, I. F., Acosta, L., *et al.* (2011). Neighborhood differences in exposure and sensitization to cockroach, mouse, dust mite, cat, and dog allergens in New York City. *The Journal of Allergy and Clinical Immunology*, 128(2), 284–292.

Peat, J. K., Tovey, E., Toelle, B. G., *et al.* (1996). House dust mite allergens: a major risk factor for childhood asthma in Australia. *American Journal of Respiratory and Critical Care Medicine*, 153(1), 141–146.

Perry, T. T., Rettiganti, M., Brown, R. H., Nick, T. G., Jones, S. M. (2012). Uncontrolled asthma and factors related to morbidity in an impoverished, rural environment. *Annals of Allergy, Asthma & Immunology*, 108(4), 254–259.

Perzanowski, M. S., Rönmark, E., Platts-Mills, T. A. E., Lundbäck, B. (2002). Effect of cat and dog ownership on sensitization and development of asthma among preteenage children. *American Journal of Respiratory and Critical Care Medicine*, 166(5), 696–702.

Perzanowski, M. S., Sporik, R., Squillace, S. P., *et al.* (1998). Association of sensitization to *Alternaria* allergens with asthma among school-age children. *The Journal of Allergy and Clinical Immunology*, 101(5), 626–632.

Phipatanakul, W., Eggleston, P. A., Wright, E. C., Wood, R. A. (2000a). Risk factors for sensitization to mouse allergen in inner-city children with asthma. *The Journal of Allergy and Clinical Immunology*, 105(1), S79.

Phipatanakul, W., Eggleston, P. A., Wright, E. C., Wood, R. A., and the National Cooperative Inner-City Asthma Study (2000b). Mouse allergen. I. The prevalence of mouse allergen in inner-city homes. *The Journal of Allergy and Clinical Immunology*, 106(6), 1070–1074.

Phipatanakul, W., Eggleston, P. A., Wright, E. C., Wood, R. A., and the National Cooperative Inner-City Asthma Study (2000c). Mouse allergen. II. The relationship of mouse allergen exposure to mouse sensitization and asthma morbidity in inner-city children with asthma. *The Journal of Allergy and Clinical Immunology*, 106(6), 1075–1080.

Phipatanakul, W., Gold, D. R., Muilenberg, M., *et al.* (2005). Predictors of indoor exposure to mouse allergen in urban and suburban homes in Boston. *Allergy*, 60(5), 697–701.

Phipatanakul, W., Matsui, E., Portnoy, J., *et al.* (2012). Environmental assessment and exposure reduction of rodents: a practice parameter. *Annals of Allergy, Asthma & Immunology*, 109(6), 375–387.

Platts-Mills, T. A. E., de Weck, A. L., Aalberse, R. C., *et al.* (1989). Dust mite allergens and asthma – a worldwide problem. *The Journal of Allergy and Clinical Immunology*, 83(2), 416–427.

Platts-Mills, T. A. E., Vervloet, D., Thomas, W. R., Aalberse, R. C., Chapman, M. D. (1997). Indoor allergens and asthma: report of the Third International Workshop. *The Journal of Allergy and Clinical Immunology*, 100(6), S2–S24.

Pongracic, J. A., O'Connor, G. T., Muilenberg, M. L., *et al.* (2010). Differential effects of outdoor versus indoor fungal spores on asthma morbidity in inner-city children. *The Journal of Allergy and Clinical Immunology*, 125(3), 593–599.

Portnoy, J., Chapman, J., Burge, H., Muilenberg, M., Solomon, W. (1987). *Epicoccum* allergy: skin reaction patterns and spore/mycelium disparities recognized by IgG and IgE ELISA inhibition. *Annals of Allergy*, 59(1), 39–43.

Portnoy, J., Chew, G. L., Phipatanakul, W., *et al.* (2013). Environmental assessment and exposure reduction of cockroaches: a practice parameter. *The Journal of Allergy and Clinical Immunology*, 132(4), 802–808, 808.e1–808.e25.

Prasad, C., Hogan, M. B., Peele, K., Wilson, N. W. (2009). Effect of evaporative coolers on skin test reactivity to dust mites and molds in a desert environment. *Allergy and Asthma Proceedings*, 30(6), 624–627.

Rabito, F. A., Iqbal, S., Kiernan, M. P., Holt, E., Chew, G. L. (2008). Children's respiratory health and mold levels in New Orleans after Katrina: a preliminary look. *The Journal of Allergy and Clinical Immunology*, 121(3), 622–625.

Rabito, F. A., Perry, S., Davis, W. E., Yau, C. L., Levetin, E. (2010). The relationship between mold exposure and allergic response in post-Katrina New Orleans. *Journal of Allergy*, 2010, 510380.

Rose, G., Arlian, L., Bernstein, D., *et al.* (1996). Evaluation of household dust mite exposure and levels of specific IgE and IgG antibodies in asthmatic patients enrolled in a trial of immunotherapy. *The Journal of Allergy and Clinical Immunology*, 97(5), 1071–1078.

Rosenfeld, L., Rudd, R., Chew, G. L., Emmons, K., Acevedo-García, D. (2010). Are neighborhood-level characteristics associated with indoor allergens in the household? *Journal of Asthma*, 47(1), 66–75.

Rosenstreich, D. L., Eggleston, P., Kattan, M., *et al.* (1997). The role of cockroach allergy and exposure to cockroach allergen in causing morbidity among inner-city children with asthma. *The New England Journal of Medicine*, 336(19), 1356–1363.

Ross, M. A., Curtis, L., Scheff, P. A., *et al.* (2000). Association of asthma symptoms and severity with indoor bioaerosols. *Allergy*, 55(8), 705–711.

Ross, M. A., Persky, V. W., Scheff, P. A., *et al.* (2002). Effect of ozone and aeroallergens on the respiratory health of asthmatics. *Archives of Environmental Health*, 57(6), 568–578.

Salo, P. M., Calatroni, A., Gergen, P. J., *et al.* (2011). Allergy-related outcomes in relation to serum IgE: results from the National Health and Nutrition Examination Survey 2005–2006. *The Journal of Allergy and Clinical Immunology*, 127(5), 1226–1235.e7.

Salo, P. M., Yin, M., Arbes Jr, S. J., *et al.* (2005). Dustborne *Alternaria alternata* antigens in US homes: results from the National Survey of Lead and Allergens in Housing. *The Journal of Allergy and Clinical Immunology*, 116(3), 623–629.

Santilli Jr, J., Rockwell, W. J., Collins, R. P. (1985). The significance of the spores of the Basidiomycetes (mushrooms and their allies) in bronchial asthma and allergic rhinitis. *Annals of Allergy*, 55(3), 469–471.

Santilli Jr, J., Rockwell, W. J., Collins, R. P. (1990). Individual patterns of immediate skin reactivity to mold extracts. *Annals of Allergy*, 65(6), 454–458.

Schumacher, M. J., Tait, B. D., Holmes, M. C. (1981). Allergy to murine antigens in a biological research institute. *The Journal of Allergy and Clinical Immunology*, 68(4), 310–318.

Sears, M. R., Herbison, G. P., Holdaway, M. D., *et al.* (1989). The relative risks of sensitivity to grass pollen, house dust mite and cat dander in the development of childhood asthma. *Clinical and Experimental Allergy*, 19(4), 419–424.

Simon-Nobbe, B., Denk, U., Pöll, V., Rid, R., Breitenbach, M. (2008). The spectrum of fungal allergy. *International Archives of Allergy and Immunology*, 145(1), 58–86.

Sprenger, J. D., Altman, L. C., O'Neil, C. E., *et al.* (1988). Prevalence of basidiospore allergy in the Pacific Northwest. *The Journal of Allergy and Clinical Immunology*, 82(6), 1076–1080.

Stelmach, I., Jerzynska, J., Stelmach, W., *et al.* (2002). Cockroach allergy and exposure to cockroach allergen in Polish children with asthma. *Allergy*, 57(8), 701–705.

Stevens, W., Addo-Yobo, E., Roper, J., *et al.* (2011). Differences in both prevalence and titre of specific immunoglobulin E among children with asthma in affluent and poor communities within a large town in Ghana. *Clinical & Experimental Allergy*, 41(11), 1587–1594.

Sundell, J., Wickman, M., Pershagen, G., Nordvall, S. L. (1995). Ventilation in homes infested by house-dust mites. *Allergy*, 50(2), 106–112.

Szánthó, A., Osváth, P., Horváth, Zs., Novák, E. K., Kujalek, É. (1992). Study of mold allergy in asthmatic children in Hungary. *Journal of Investigational Allergology and Clinical Immunology*, 2(2), 84–90.

Tovey, E. R., Chapman, M. D., Platts-Mills, T. A. E. (1981). Mite faeces are a major source of house dust allergens. *Nature*, 289(5798), 592–593.

Tovey, E., DeLucca, S., Pavlicek, P., *et al.* (2000). The morphology of particles carrying mite, dog, cockroach, and cat aeroallergens affects their efficiency of collection by nasal samplers and cascade impactors. *The Journal of Allergy and Clinical Immunology*, 105(1), S228.

van Strien, R. T., Verhoeff, A. P., Brunekreef, B., van Wijnen, J. H. (1994). Mite antigen in house dust: relationship with different housing characteristics in the Netherlands. *Clinical and Experimental Allergy*, 24(9), 843–853.

Venables, K. M., Tee, R. D., Hawkins, E. R., *et al.* (1988). Laboratory animal allergy in a pharmaceutical company. *British Journal of Industrial Medicine*, 45(10), 660–666.

Vijay, H. M., Kurup, V. P. (2004). Fungal allergens. *Clinical Allergy and Immunology*, 18, 223–249.

Wahn, U., Lau, S., Bergmann, R., *et al.* (1997). Indoor allergen exposure is a risk factor for sensitization during the first three years of life. *The Journal of Allergy and Clinical Immunology*, 99(6), 763–769.

Wålinder, R., Ernstgård, L., Johanson, G., *et al.* (2005). Acute effects of a fungal volatile compound. *Environmental Health Perspectives*, 113(12), 1775–1778.

Wickens, K., Mason, K., Fitzharris, P., *et al.* (2001). The importance of housing characteristics in determining Der p 1 levels in carpets in New Zealand homes. *Clinical and Experimental Allergy*, 31(6), 827–835.

Wood, R. A., Eggleston, P. A., Lind, P., *et al.* (1988). Antigenic analysis of household dust samples. *American Review of Respiratory Disease*, 137(2), 358–363.

Zureik, M., Neukirch, C., Leynaert, B., *et al.* (2002). Sensitisation to airborne moulds and severity of asthma: cross sectional study from European Community respiratory health survey. *British Medical Journal*, 325(7361), 411–417.

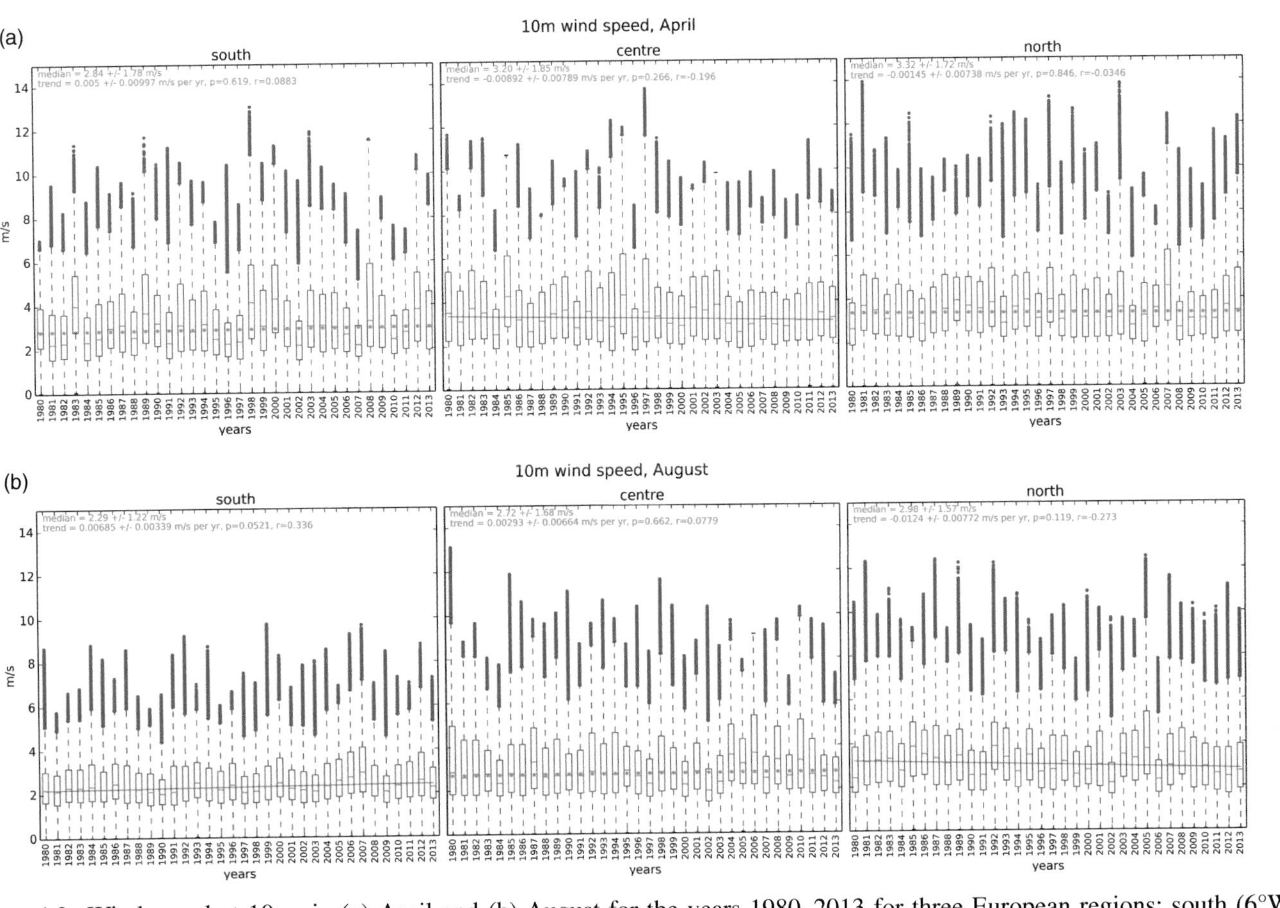

Figure 4.2. Wind speed at 10 m in (a) April and (b) August for the years 1980–2013 for three European regions: south (6°W, 38°N – 3°W, 41°N), central (10°E, 49°N – 13°E, 52°N), north (22°E, 61°N – 25°E, 64°N). Median and its trend are marked by red; data quartiles and outliers are blue. (Unit: m s^{-1}.)

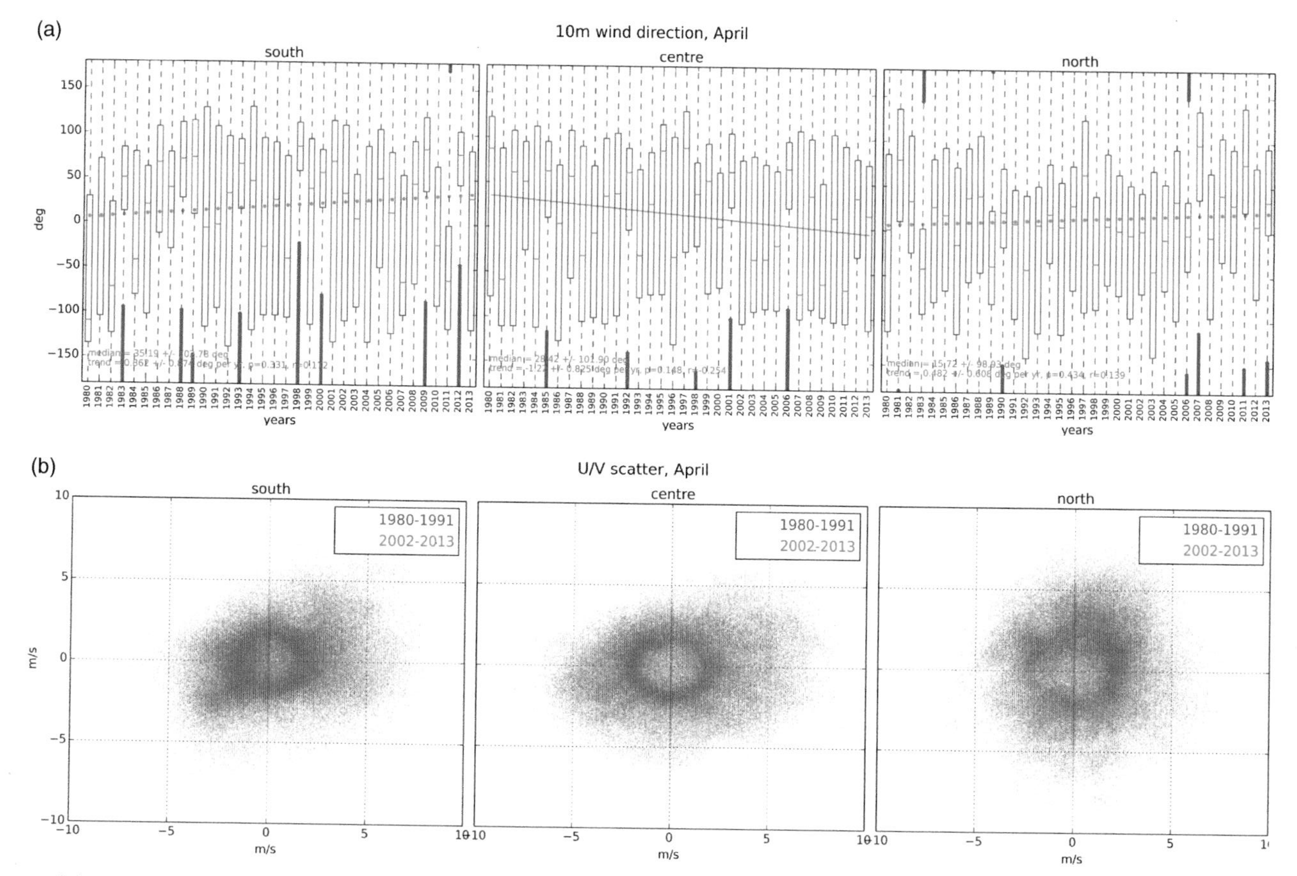

Figure 4.3. Wind direction at 10 m (ϕ_{10}) in April for the years 1980–2013 (a), and u_{10}–v_{10} scatter plots for wind at 10 m (b). Regions and notations are the same as in Figure 4.2. (Unit: degrees for (a) and m s^{-1} for (b).)

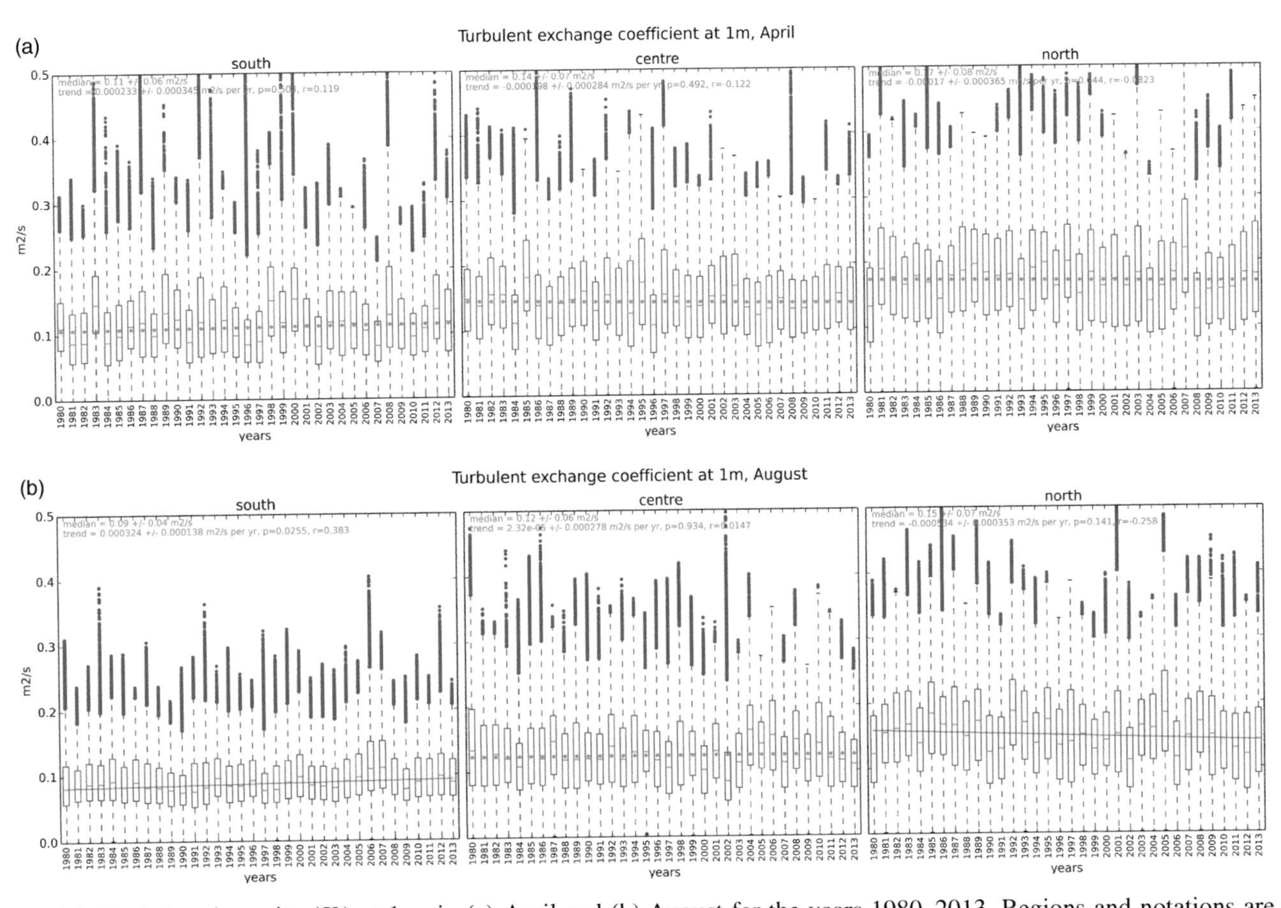

Figure 4.4. Turbulent intensity (K_z) at 1 m in (a) April and (b) August for the years 1980–2013. Regions and notations are the same as in Figure 4.2. (Unit: $m^2\ s^{-1}$.)

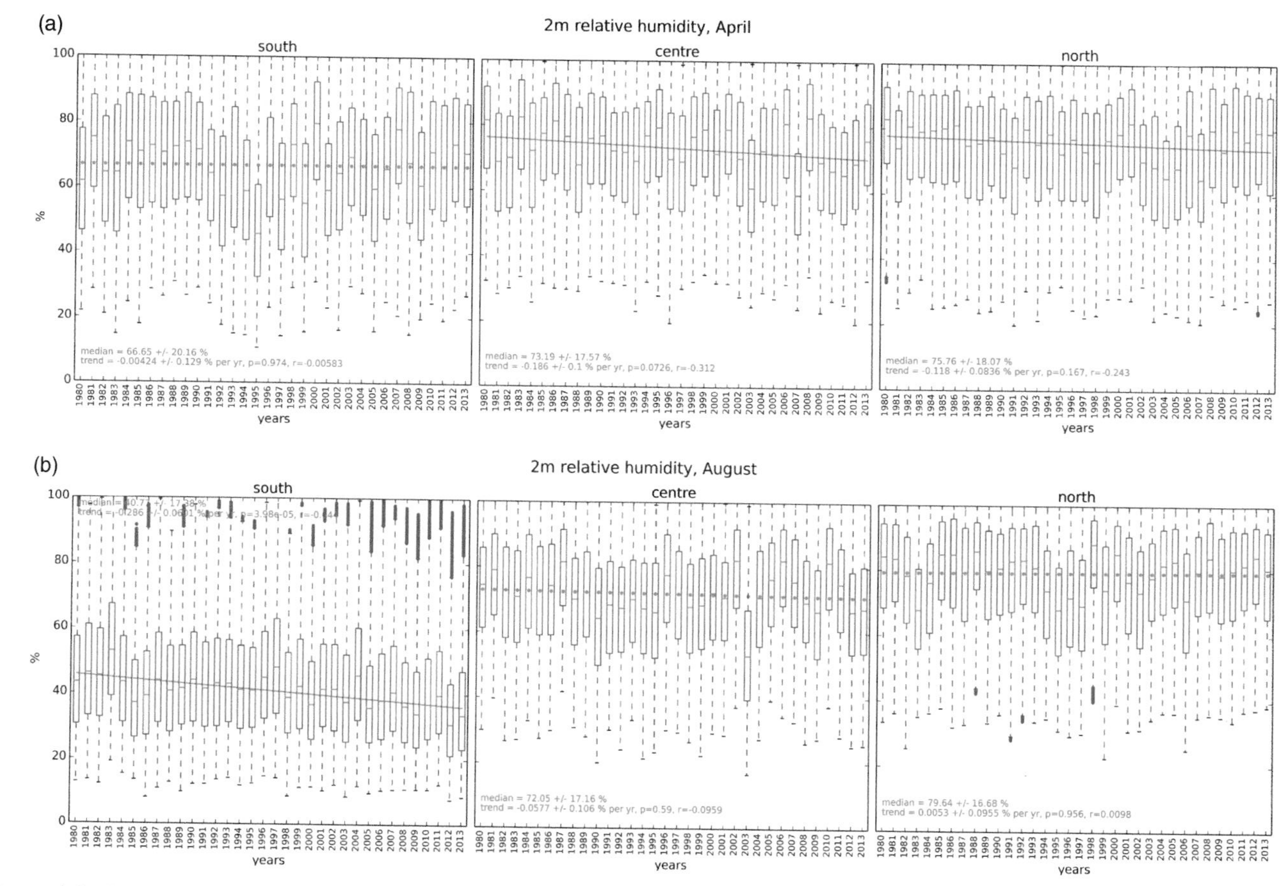

Figure 4.5. Relative humidity at 2 m in (a) April and (b) August for the years 1980–2013. Regions and notations are the same as in Figure 4.2. (Unit: percentage.)

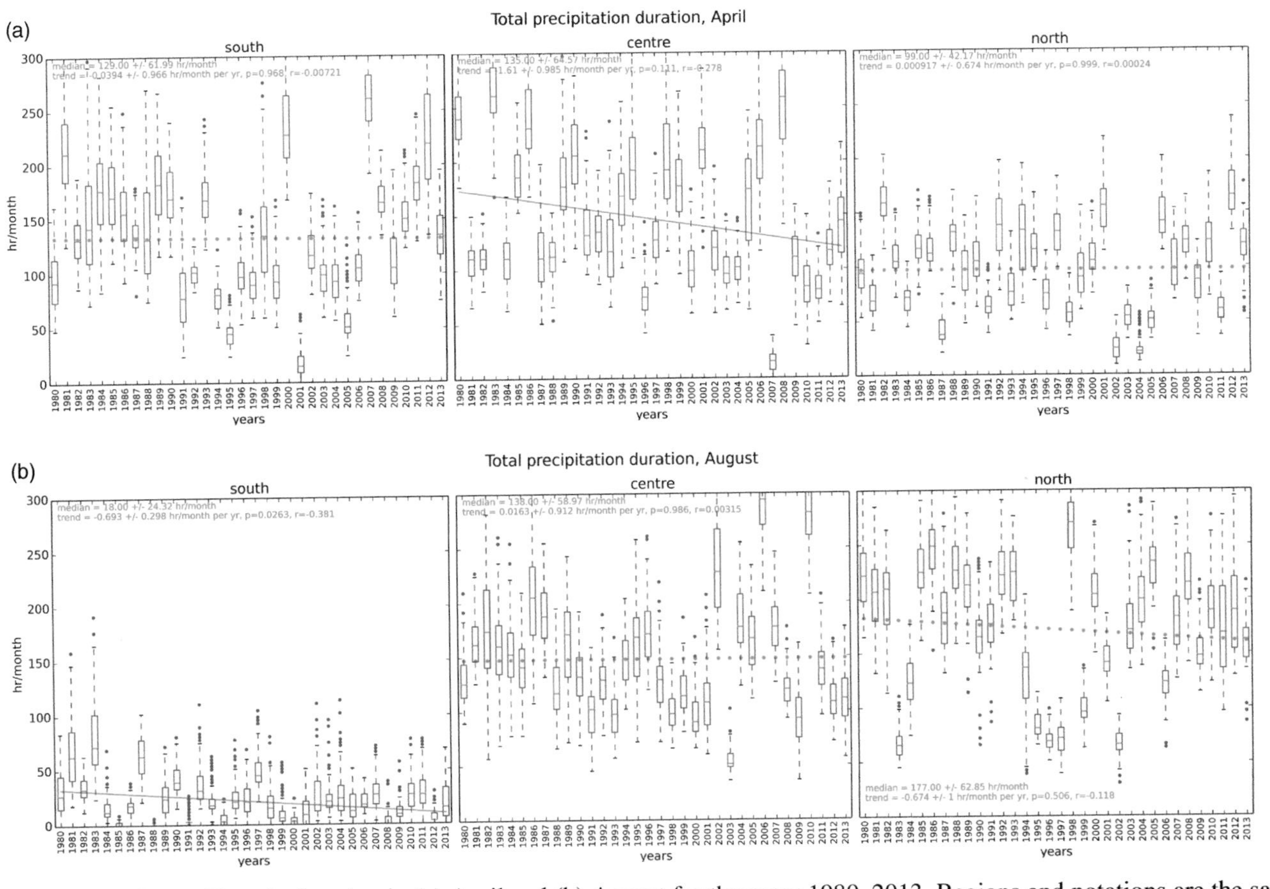

Figure 4.6. Total monthly rain duration in (a) April and (b) August for the years 1980–2013. Regions and notations are the same as in Figure 4.2. (Unit: hour.)

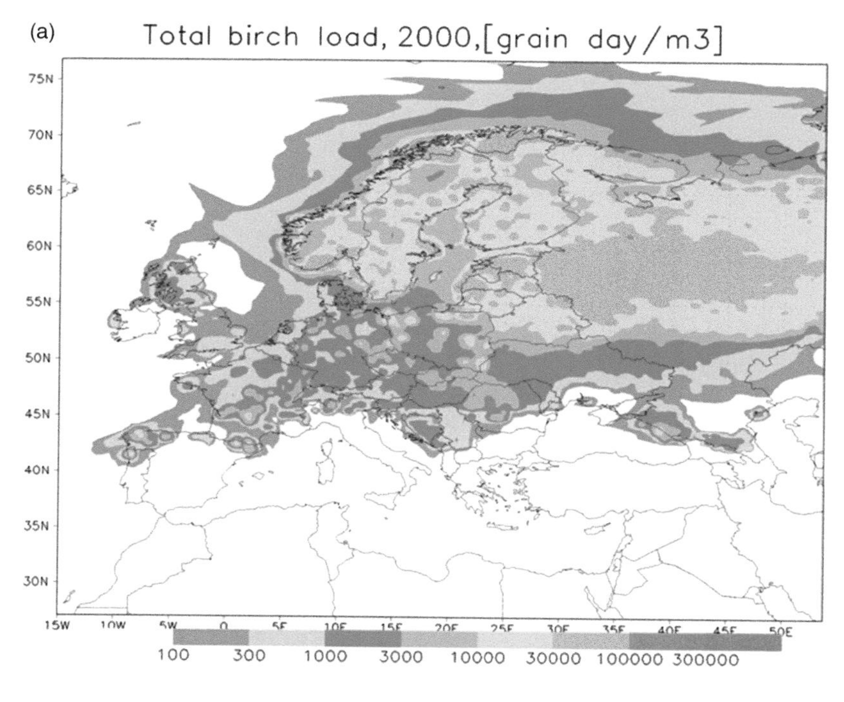

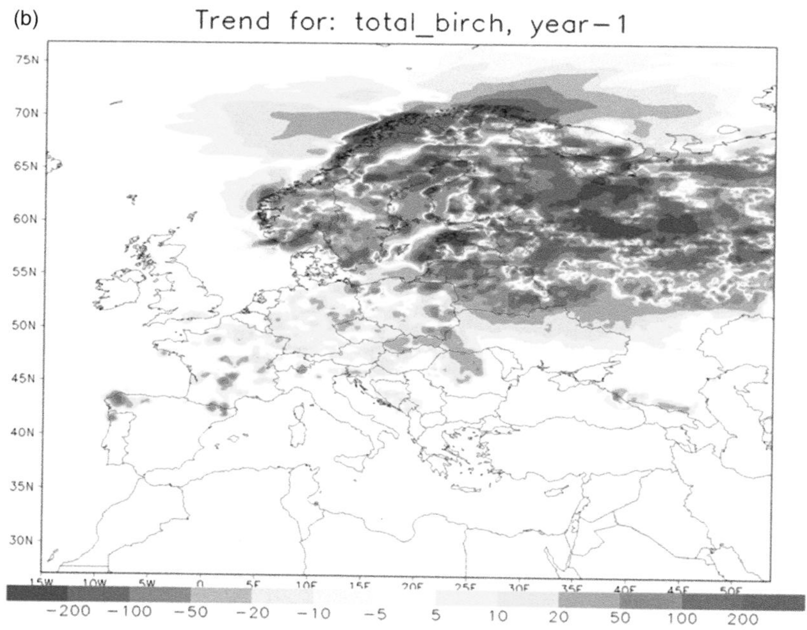

Figure 4.7. Birch total seasonal pollen count in Europe for 2000 (a; pollen day m^{-3}) and its 1980–2012 trend (b; pollen day m^{-3} yr^{-1}).

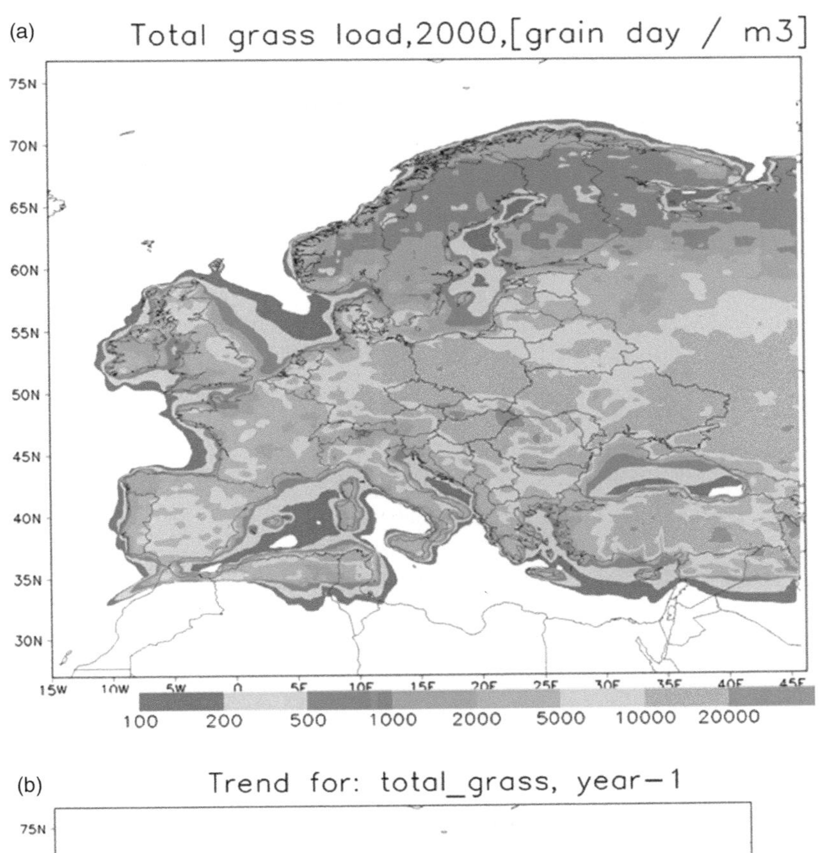

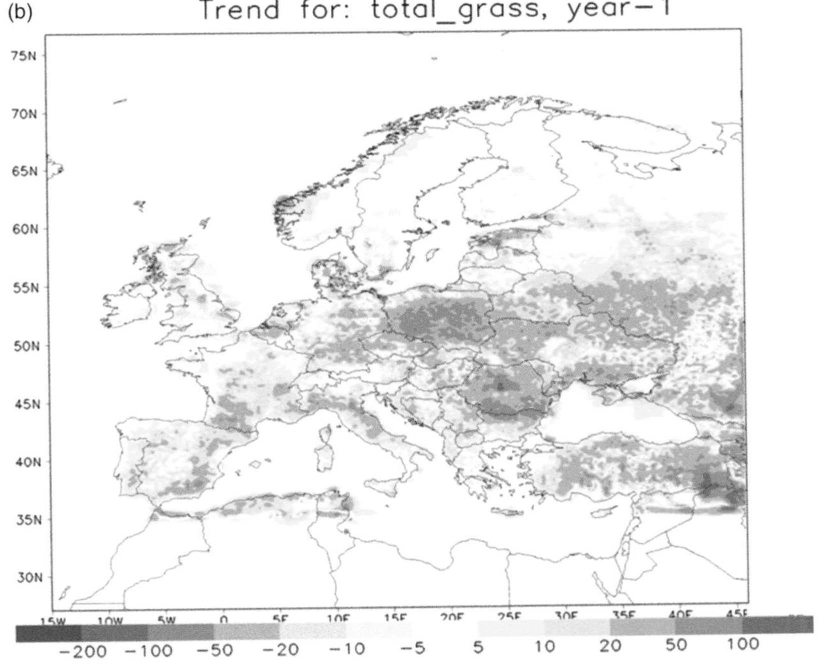

Figure 4.8. Grass total seasonal pollen count in Europe for 2000 (a; pollen day m^{-3}) and its 1980–2012 trend (b; pollen day m^{-3} yr^{-1}).

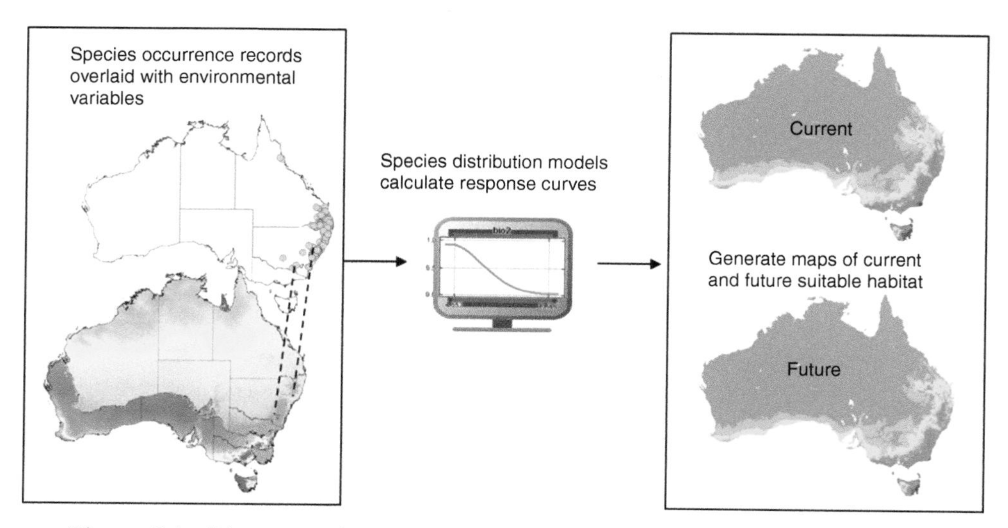

Figure 3.1. Diagrammatic representation of a correlative species distribution model. Environmental characteristics of locations where species occur are used to generate models of species–environment relationships, which can then be projected onto scenarios of climate (past, current, or future) to identify potential distributions.

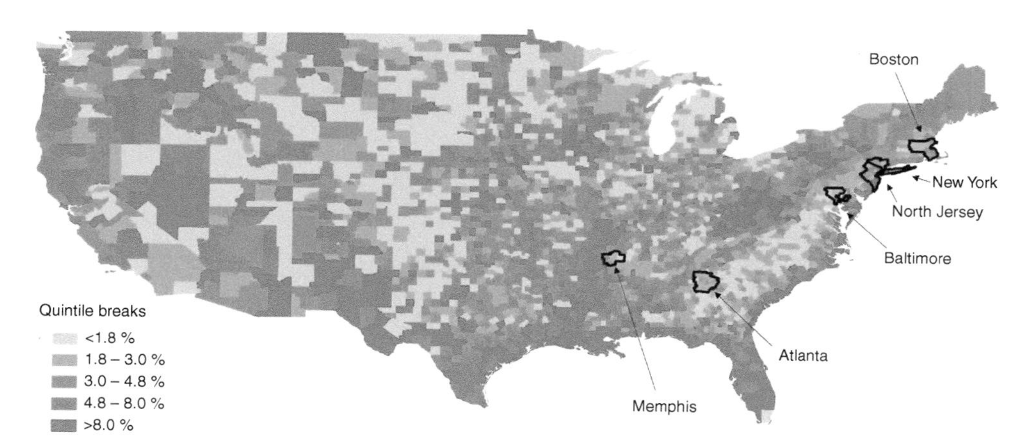

Figure 7.6. US map representing percentage of housing units in each county in 100-year flood hazard areas (data obtained from FEMA: hazards.fema.gov/femaportal/wps/portal/NFHLWMS). The colour codes represent quintiles calculated based on the distribution of percentage of housing units in the flood hazard area. Counties shaded with darker red indicate a higher proportion of housing units that are in flood hazard areas compared to those in lighter red which indicate a relatively lower percentage of housing units in flood hazard areas. The map also shows the six MSAs selected based on availability of housing structure information from the AHS.

8

Interactions among Climate Change, Air Pollutants, and Aeroallergens

PATRICK L. KINNEY[1], KATE R. WEINBERGER[2], AND RACHEL L. MILLER[3]

[1]*Climate and Health Program*
Mailman School of Public Health
Columbia University
[2]*Institute at Brown for Environment & Society*
Brown University
[3]*Division of Pulmonary, Allergy, and Critical Care Medicine*
Columbia University Medical Center

8.1 Introduction

Anthropogenic air pollution represents an important class of air contaminants with well-known respiratory health effects in humans. Because of the ubiquity of air pollution in both indoor and outdoor environments, it is reasonable to assume that exposures to aeroallergens rarely occur without co-exposure to one or more air pollutants. This highlights the importance of possible interactions between air pollution and aeroallergens. Such interactions may occur over various scales of space and time, involving mechanisms at the level of individual particles in the atmosphere, via patterns of deposition in the respiratory system, and at the level of the individual immune responses to environmental exposures. Furthermore, air pollution emission, transport, chemical transformation, dispersion, and deposition all are influenced by meteorological factors that are undergoing change in response to climate change. For example, a warming climate will favour higher concentrations of ozone in some locations. On the other hand, actions to mitigate climate change can have profoundly positive impacts on local air quality, leading to substantial health benefits.

Here we examine the interactions of air pollution and aeroallergens in the context of a changing climate. We start by reviewing emerging knowledge on how climate change affects air pollution. Then, after briefly reviewing the health effects of air pollution, we examine the small but growing literature on interactions between air pollution and aeroallergens at the levels of exposure, dose, and health effects.

8.2 Impact of Climate Change on Air Pollution

The influence of meteorology on air quality is substantial and well established (Jacob, 2005; Jacob and Winner, 2009; Kinney, 2008). Weather patterns have a profound influence on the concentration of air pollution in the atmosphere through the action of temperature, winds, vertical mixing, and rainfall. Air pollution episodes often occur when atmospheric conditions reduce vertical and/or horizontal dispersion. For example, calm winds and warm air aloft create inversions that limit the vertical dispersion of ground-level pollution emissions in cities. Stable high-pressure systems can allow pollution to accumulate and intensify over large regions. Emissions from power plants increase during heat waves when air-conditioning use peaks. Also, cold temperatures lead to higher emissions from residential fuel combustion (Jung *et al.*, 2010).

Future changes in climate likely will lead to changing patterns of air pollution. Higher temperatures hasten the chemical reactions that lead to ozone and secondary particle formation (Likhvar *et al.*, 2015; Sujaritpong *et al.*, 2014; Weaver *et al.*, 2009). Higher temperatures, and perhaps elevated carbon dioxide (CO_2) concentrations, also lead to increased emissions of ozone-relevant volatile organic compound precursors (Hogrefe *et al.*, 2005). On the other hand, pollution emissions from winter fuel combustion for heating of buildings will likely decrease.

Understanding how these and other climate-related factors will affect future air pollution in specific regions and locales is challenging, due in part to the uncertainties in modelling these relationships and also the strong influence of future air pollution emissions, independent of climate-related influences (Hedegaard *et al.*, 2013).

Atmospheric climate and chemistry models differ considerably with respect to future projections of small-scale patterns of the key meteorological and chemical processes, leading to wide variations in their future projections of air pollution. Furthermore, some key atmospheric chemistry processes, such as nitrate and secondary organic aerosol formation, remain uncertain or are not fully included in such models. As a result, future air pollution projections from one or a few models must be viewed with extreme scepticism. It is preferable to base projections on an ensemble of multiple models.

Over the past 10 years, a modelling literature examining future climate influences on air quality has emerged and rapidly expanded (Air Quality Expert Group, 2007; Andersson and Engardt, 2010; Carvalho *et al.*, 2010; Cecchi *et al.*, 2010; Chang *et al.*, 2010; Dawson *et al.*, 2007a, 2007b, 2008; Forkel and Knoche, 2006; Gao *et al.*, 2013; Hedegaard *et al.*, 2013; Hogrefe *et al.*, 2004; Horton *et al.*, 2012; Huang *et al.*, 2007; Katragkou *et al.*, 2011; Lam *et al.*, 2011; Langner *et al.*, 2012; Leung and Gustafson Jr, 2005; Liang *et al.*, 2006; Liao *et al.*, 2010; Likhvar *et al.*, 2015; Mahmud *et al.*, 2012; Mickley *et al.*, 2004; Murazaki and Hess, 2006; Nolte *et al.*, 2008; Racherla and Adams, 2006, 2008, 2009; Steiner *et al.*, 2006; Stevenson *et al.*, 2006; Tagaris *et al.*, 2007, 2008; Tai *et al.*, 2012; Tao *et al.*, 2007; The Royal Society, 2008; Weaver *et al.*, 2009; Wu *et al.*, 2008; Zeng and Pyle, 2003). A smaller but growing literature has further examined potential human health impacts of future air pollution under climate change scenarios (Chang *et al.*, 2010; Fang *et al.*, 2013a, 2013b; Gosling *et al.*, 2007, 2012; Jacobson, 2008; Kan *et al.*, 2012; Knowlton *et al.*, 2004, 2008; Likhvar *et al.*, 2015; Orru *et al.*, 2013; Post *et al.*, 2012; Sheffield *et al.*, 2011; Tagaris *et al.*, 2009, 2010). Most often, studies link climate and chemistry models to simulate surface ozone (and occasionally particulate matter, $PM_{2.5}$) concentrations under one or more scenarios of future climate change and air pollution emissions.

For ozone, findings suggest that climate change alone (absent changes in pollution precursor emissions) could lead to increases in ozone episodes in polluted areas and decreases in more rural areas less influenced by anthropogenic emissions. At the regional level, results vary considerably depending on the models and assumptions used.

For particulate matter <2.5 μm in aerodynamic diameter ($PM_{2.5}$), the literature is smaller and less robust than for ozone but suggests that climate change could lead to either increases or decreases in $PM_{2.5}$ in the future, depending on location and modelling assumptions. The uncertain sensitivity to climate change in the case of $PM_{2.5}$ reflects in part the complexity of the dependence of different $PM_{2.5}$ components on meteorological variables, and in part on the coupling of aerosols to the hydrological cycle which is not well represented in general circulation models (Racherla and Adams, 2006).

For both ozone and $PM_{2.5}$, when models take into account reduced precursor emissions according to planned or projected policies in developed countries, concentrations generally come down in the future, in spite of climate change. Much less studied are climate/air pollution impacts in developing countries, where local air quality measurement and management programs are often absent.

8.3 Health Effects of Air Pollution

A very extensive body of scientific research has clarified the health effects of common community air pollutants, including ozone, particulate matter, nitrogen dioxide (NO_2), sulphur dioxide (SO_2), and others (World Health Organization, 2013). Though cardiopulmonary mortality is the most significant and costly health outcome that has been documented, more relevant for the current review are studies that have documented respiratory effects in community settings. Ozone and $PM_{2.5}$ (especially traffic-related particles including diesel exhaust particles [DEP]) are of primary interest here due to their potent health effects at ambient levels, but NO_2, SO_2, and others can also be risk factors for respiratory effects in hotspots with high concentrations (Balmes *et al.*, 1987; Bauer *et al.*, 1986).

Most anthropogenic air pollution arises due to combustion of fuels (including wood, coal, oil, gasoline and diesel, liquefied petroleum gas, and natural gas) to produce energy for transportation, heating, cooling and lighting buildings, and for waste processing and disposal. In addition, solid fuels (e.g., dung, wood, charcoal, coal) are often used for cooking or heating in developing countries. Fuel combustion produces a wide spectrum of solid and gaseous pollutants that can impact community air quality. Primary pollutants are emitted directly at the source and thus have the potential for very local impacts, whereas secondary pollutants form via reactions of primary pollutants in the atmosphere and thus have the potential for downwind regional impacts. Residents of urban areas typically are exposed to a complex mixture of both primary and secondary pollutants.

Primary pollutants include SO_2, NO_2, carbon monoxide (CO), and some components of coarse PM (PM_{10}) and $PM_{2.5}$ such as elemental and organic carbon, as well as many other less familiar or less measured pollutants such as trace metals and a wide range of solid and vapour phase organic compounds. Secondary pollutants include ozone, some components of PM_{10} and $PM_{2.5}$, and organic vapours. Among secondary particles, sulphates, nitrates, and secondary organic carbon represent the dominant components in most cases. Both PM_{10} and $PM_{2.5}$ are generic particle classes defined by the upper limit of the particle aerodynamic diameters contained within that class (i.e., 10 or 2.5 μm). These metrics encompass particles of a wide range of chemical composition and aerodynamic diameters.

Available knowledge regarding health effects of air pollution is most robust for the handful of pollutants for which routine air monitoring data are available, including ozone, NO_2, SO_2, CO, PM_{10}, and $PM_{2.5}$ (Künzli *et al.*, 2010; WHO, 2013). These are the main pollutants that have been studied epidemiologically over time and/or space in cities. Additional health knowledge for these and other pollutants is available from controlled experimental studies involving animals, cell systems, and human volunteers.

Early studies showed that ambient-level ozone exposures cause acute, reversible drops in lung volumes, increases in non-specific bronchial responsiveness, and pulmonary inflammation in humans (Devlin *et al.*, 1991; Horstman *et al.*, 1990; Kinney *et al.*, 1989, 1996). Epidemiologic studies further demonstrated associations with asthma exacerbations, emergency room visits, hospital admissions, and deaths (Bell *et al.*, 2004; Burnett *et al.*, 1997; Kinney and Özkaynak, 1991; Lin *et al.*, 2008). Populations most at risk of ozone effects include children and adults who are active outdoors (Kinney *et al.*, 1996), the elderly (Bell *et al.*, 2014), and asthmatics. Ozone's inflammatory action in small airways provides a possible mechanistic link to the asthma phenotype and to more general systemic effects (Devlin *et al.*, 1991; Kahle *et al.*, 2015). Some epidemiologic studies have investigated the independent and/or interacting effects of ozone and aeroallergens. In a panel study of twelve asthmatics, both ozone and fungal spore counts were associated with respiratory symptoms and as-needed inhaler use (Delfino *et al.*,

1996). Recent work has highlighted health effect interactions between ozone and temperature and/or season both in the United States and elsewhere (Chen *et al.*, 2014, 2015; Jhun *et al.*, 2014; Li *et al.*, 2015).

Epidemiologic studies addressing health effects of PM_{10} and $PM_{2.5}$ have reported associations with a wide range of adverse health outcomes, including short- and long-term mortality, hospitalisations, respiratory symptoms, and decreases in lung function (Pope III and Dockery, 2006). Populations at greatest risk of PM effects include the elderly, those with preexisting cardiopulmonary disease, and the very young. For example, emergency room visits, hospital admissions, and death from cardiac and cardiopulmonary disease among the elderly have been associated with higher $PM_{2.5}$ and ozone levels (Andersen *et al.*, 2012; Bentayeb *et al.*, 2012; Halonen *et al.*, 2009; Winquist *et al.*, 2012). In addition, using daily time series analyses in New York City hospitals, children had a greater risk for intensive care unit admissions for asthma for each 12 $\mu g\ m^{-3}$ increase in $PM_{2.5}$ and for each 22 ppb increase in ozone, compared to adults (Silverman and Ito, 2010). Also, the prenatal time window of exposure may be a time of heightened risk. In mice, prenatal exposure to residual oil fly ash aerosol induced greater airway hyperreactivity and allergic inflammation in the airways in sensitised offspring (Hamada *et al.*, 2007). Our group at the Columbia Center for Children's Environmental Health (CCCEH) showed that prenatal exposure to polycyclic aromatic hydrocarbons (PAHs), either in association with exposure to second-hand smoke (Miller *et al.*, 2004; Rosa *et al.*, 2011) or in association with higher cockroach allergen levels in residential dust (Perzanowski *et al.*, 2013), was associated with more respiratory symptoms and with the development of proallergic cockroach immunoglobulin E (IgE), respectively. It is not yet clear from the available evidence whether specific chemical components of PM are more important than others in the observed health effects.

Because many air pollutants covary in space and time, any one pollutant can often be thought of as a proxy for pollutants that share similar sources and atmospheric dispersion characteristics. For example, using NO_2 as a proxy for residential exposures to traffic-related air pollution, Jerrett and colleagues found associations with new-onset asthma in the Southern California Children's Health Study (Jerrett *et al.*, 2008). In the case of the semivolatile PAH pyrene, the CCCEH observed that repeated high exposure to pyrene was associated with asthma, asthma medication use, and emergency room visits for asthma (Jung *et al.*, 2012). Indicators of sources of traffic and stationary air pollution based on geographic information systems (GIS) also have been helpful in estimating personal air pollution exposure and risk for asthma-related outcomes without relying on expensive measures of many air pollutants. In another study from the CCCEH cohort, spatial data on the proximity and density of roadways and built environment were collected in a buffer surrounding children's homes and analysed longitudinally. Density of four-way intersections and percent commercial areas were associated with childhood wheeze, and proximity to stationary sources of air pollution was associated with report of asthma through age 5 years (Patel *et al.*, 2011).

Experimental studies have contributed key knowledge on interactions between air pollution and allergens including information on the dose–response relationship and potential biological mechanisms. These studies have largely focussed on DEP, an important fine particle fraction that is widely observed in urban air. In the following section, we review the extensive literature on the adjuvant effects of DEP.

8.4 Air Pollution and Pollen Host Interactions (Adjuvant Effects)

In the previous section, we reviewed some of the ways that air pollutants can affect human health. Here we consider the specific question of how air pollution may interact with allergens in causing adverse health effects. Much of the relevant literature has focussed on diesel exhaust particulate matter. Evidence has shown that DEP can act in a synergistic manner with allergens such as pollen to enhance the development of allergic disease. DEP is an important component of ambient $PM_{2.5}$ in many locations, composed of a non-extractable elemental carbon core with adsorbed organic chemicals such as PAHs (Westerholm *et al.*, 1991). Approximately 95% of DEP is <1 μm in aerodynamic diameter, making it easily respirable (McClellan, 1987).

The first in vivo studies demonstrating a synergistic effect of DEP and allergen date back to the 1986 work of Muranaka *et al.* (Muranaka *et al.*, 1986), who found that mice immunised with both DEP and ovalbumin, an antigen commonly used in mouse models of asthma to induce allergic sensitisation, produced higher levels of the allergic antibodies, ovalbumin-specific IgE, than mice that received only ovalbumin. This observation suggested that DEP acted as an adjuvant to facilitate the process of specific allergic sensitisation. This finding has been replicated numerous times over the last three decades (Fujimaki *et al.*, 1994; Heo *et al.*, 2001; Løvik *et al.*, 1997; Nygaard *et al.*, 2004; Samuelsen *et al.*, 2008; Takafuji *et al.*, 1987), with remarkably consistent results across different ovalbumin and DEP administration routes (Fujimaki *et al.*, 1994; Muranaka *et al.*, 1986; Takafuji *et al.*, 1987) and different animal models (Kobayashi, 2000; Steerenberg *et al.*, 1999). The adjuvant effect of DEP has also been demonstrated for allergens other than ovalbumin. Combined exposure to DEP and other allergens such as house dust mite (Kadkhoda *et al.*, 2005; Suzuki *et al.*, 1996), several types of pollen including Japanese cedar (*Cryptomeria japonica*), timothy grass (*Phleum pratense*), and birch (*Betula* sp.), and the mould *Aspergillus fumigatus* (Fernvik *et al.*, 2002; Fujimaki *et al.*, 1994; Kanoh *et al.*, 1996; Liu *et al.*, 2013; Maejima *et al.*, 1997; Steerenberg *et al.*, 1999) leads to significantly increased total and/or allergen-specific IgE levels compared to exposure to allergen alone. In addition to its effect on IgE, DEP also enhances eosinophilic airway inflammation, airway hyperresponsiveness, and airway remodelling in mouse models of asthma (Matsumoto *et al.*, 2006; Takahashi *et al.*, 2010).

Despite the consistency of these findings, the relative importance of the different components of DEP in promoting allergen-specific IgE production is an area of active debate. Various PAHs, which are often bound to the surface of DEP, enhance allergen-specific IgE production when administered with allergen (Heo *et al.*, 2001; Kadkhoda *et al.*, 2005; Suzuki *et al.*, 1996), suggesting that the bound organic components are responsible for the adjuvant effect. However, other studies using polystyrene particles to mimic the DEP core have shown similar effects (Granum *et al.*, 2001; Nygaard *et al.*, 2004), indicating that the physical properties of the core itself may be important as well. Furthermore, other types of particulate matter (e.g., carbon black, fly ash, wood smoke, road traffic particles) also enhance allergen-specific IgE production; however, the adjuvant effect appears to be strongest for DEP (Løvik *et al.*, 1997; Maejima *et al.*, 1997; Samuelsen *et al.*, 2008).

The large body of work conducted in animal models is supported by several human experimental studies. In a seminal experiment, Diaz-Sanchez *et al.* (1999) demonstrated that DEP can facilitate allergic sensitisation to a novel allergen. In this study, human subjects were exposed with or without DEP to the novel allergen keyhole limpet hemocyanin (KLH). Over half the subjects exposed to both KLH and DEP developed specific IgE against KLH, compared to none among subjects who were exposed only to KLH (Diaz-Sanchez *et al.*, 1999). Experiments in humans also have shown that DEP can enhance specific IgE production in individuals who are already sensitised to an allergen. In subjects sensitised to the major ragweed (*Ambrosia* sp.) allergen Amb a 1, Diaz-Sanchez *et al.* found that nasal challenge with DEP and Amb a 1 resulted in a sixteen-fold increase in Amb a 1–specific IgE compared to challenge with Amb a 1 alone, suggesting that DEP can enhance allergic disease at multiple steps in the atopic march (Diaz-Sanchez *et al.*, 1997).

A variety of mechanisms have been proposed to explain the adjuvant effect of DEP. While DEP can act directly on B-cells to increase ongoing IgE production in the absence of allergen in humans (Takenaka *et al.*, 1995), it may act with allergen on the cytokine milieu to increase specific IgE production. Both animal and human studies show that combined exposure to DEP and allergen increases levels of type-2 T-helper cell (Th2) cytokines such as interleukin 4 (IL-4), which promote the development of IgE-dependent allergic sensitisation and disease, as well as decreases levels of type-1 T-helper cell (Th1) cytokines such as interferon gamma (IFN-γ) which protect against an allergic phenotype (Diaz-Sanchez *et al.*, 1996, 1997; Fujimaki *et al.*, 1994; Nygaard *et al.*, 2005; Samuelsen *et al.*, 2008). One study found that changes in the cytokine environment could be driven by epigenetic changes in response to DEP and allergen exposure. Specifically, exposure to both DEP and *Aspergillus fumigatus* resulted in hypomethylation of the IL-4 promoter and hypermethylation of the IFN-γ promoter in mice that correlated with IgE levels (Liu *et al.*, 2008). Measures of DEP estimated by land use regression analyses also were associated in the Cincinnati Childhood Allergy and Air Pollution Study with increased DNA methylation of Foxp3 (Brunst *et al.*, 2013), the T regulatory transcription factor whose suppression has been implicated in asthma-related outcomes.

DEP also may contribute to the activation and maturation of dendritic cells, which are critical initiators of the immune response to allergen (Yoshida *et al.*, 2010). Others have postulated that the adjuvant effect may be the result of DEP binding to and delivering allergen to the lower airways (Ormstad, 2000). Finally, DEP can disrupt the epithelial barrier of the airway, allowing allergen to come into contact with antigen-presenting cells such as dendritic cells (Devalia *et al.*, 1996). However, DEP increases specific IgE production even when it is injected rather than inhaled (Løvik *et al.*, 1997), suggesting that other mechanisms are involved. Additional mechanisms may include alteration of T regulatory cell function (Liu *et al.*, 2013; Nadeau *et al.*, 2010), impairment of β2 adrenergic receptor signalling (Chu *et al.*, 2013; Factor *et al.*, 2011; Fu *et al.*, 2012), and greater production of fibrinogen and other coagulatory disturbances (Bind *et al.*, 2012; Inoue *et al.*, 2006). This topic has been reviewed recently by Miller and Peden (2014).

Epidemiologic studies also provide evidence of diesel–allergen interactions in human populations, although the results have been less clear-cut than in animal studies. Some of the earliest evidence comes from studies that date back several decades, which showed that populations living in areas with higher levels of traffic or air pollution were more likely to be sensitised to inhalant allergens (Heinrich *et al.*, 1999; Ishizaki *et al.*, 1987). More recent studies that have collected exposure and outcome information at the individual level show that children who are exposed to higher levels of PAHs are more likely to be sensitised to indoor allergens such as mouse and cockroach (Miller *et al.*, 2010; Perzanowski *et al.*, 2013). In a large European cohort, children with higher exposure to $PM_{2.5}$ and who lived closer to major roads were more likely to develop sensitisation to inhalant allergens (Morgenstern *et al.*, 2008). Patel and colleagues, in their prospective analysis of GIS data, also showed that an increase in proximity to highways and a greater percentage of commercial areas were each associated with increased IgE levels (Patel *et al.*, 2011). However, a subsequent meta-analysis of five European cohorts including this one found no association between estimated $PM_{2.5}$ exposure and allergic sensitisation (Gruzieva *et al.*, 2014). While DEP is an important component of $PM_{2.5}$, this null finding could be the result of different exposures to DEP itself across cohorts.

One additional explanation for the mixed results observed in human populations may have to do with differential susceptibility to the adjuvant effect of DEP. Glutathione-S-transferases are a group of proteins that are involved in the detoxification of xenobiotics including many of the chemicals bound to the DEP core (Prestera and Talalay, 1995). In a randomised controlled crossover trial of human subjects who were sensitised to ragweed, Gilliland *et al.* found that subjects who had either the null genotype for the glutathione-S-transferase gene GSTM1 or the I105 genotype for the GSTP1 gene produced significantly more ragweed-specific IgE and histamine when challenged with combined ragweed and DEP compared to subjects with other genotypes (Gilliland *et al.*, 2004). Similarly, in an epidemiologic study of early-life exposure to allergens, Perzanowski *et al.* found that the association between PAH exposure and sensitisation to cockroach allergen was strongest in children who had the null GSTM1 genotype (Perzanowski *et al.*, 2013). These results indicate that individuals who metabolise DEP less efficiently may be more sensitive to its adjuvant effects.

8.5 Air Pollution and Pollen Interactions in the Atmosphere

Air pollutants can interact with pollen and pollen-producing plants even before they enter the human body. While most whole pollen grains have large diameters (Driessen and Quanjer, 1991), respirable allergenic proteins and glycoproteins contained within the pollen grain are released upon contact with water (Howlett and Knox, 1984). Several laboratory studies demonstrate that these proteins can bind to DEP in the atmosphere. Such complexes can penetrate into the lower airways, resulting in simultaneous co-exposure to DEP and allergen. For example, the grass pollen allergen Lol p 1 binds to DEP in a laboratory setting (Knox *et al.*, 1997), as does the birch pollen allergen Bet v 1 (Steinsvik *et al.*, 1998). Similarly, Ormstad *et al.* found that DEP binds to several major indoor and outdoor allergens, including Bet v 1, Fel d 1 (cat allergen), and Can f 1 (dog allergen) in vitro (Ormstad *et al.*, 1998). Furthermore, these same three allergens were bound to indoor suspended particulate matter samples from houses in Oslo, Norway, suggesting that allergen–DEP binding occurs in real-world settings (Ormstad, 2000). Similarly, Namork *et al.* found pollen allergens bound to combustion particles sampled from four European cities (Namork *et al.*, 2006).

In addition, there is a growing body of literature demonstrating that air pollutants can enhance the allergenicity of pollen grains. For example, Ghiani *et al.*

(2012) sampled ragweed pollen from high-traffic, low-traffic, and vegetated areas in the Po River plain in Italy and found that pollen from the high-traffic areas was more allergenic than pollen from the low-traffic and vegetated areas. This higher allergenicity was driven both by the presence of larger amounts of the major ragweed allergens Amb a 1 and Amb a 2, as well as by the increased expression of the minor allergens Amb a 6 and Amb a 10 (Ghiani *et al.*, 2012). Post-translational modifications to allergens affecting IgE recognition have also been proposed as a mechanism by which air pollutants either increase or decrease pollen allergenicity (Ghiani *et al.*, 2012; Petersen *et al.*, 1998; Rogerieux *et al.*, 2007).

Similar increases in allergen content have been documented in other species. For example, pollen from *Cupressus arizonica* collected from more heavily polluted regions contain more of the allergen Cup a 3 than pollen collected from less polluted regions in Spain. This may occur as a defence against environmental stressors such as air pollutants (Cortegano *et al.*, 2004; Suárez-Cervera *et al.*, 2008). In another field study conducted in a single urban environment, the amount of allergen Bet v 1 contained in birch pollen grains was higher in areas that had higher ozone concentrations (Beck *et al.*, 2013). In a laboratory setting, Masuch *et al.* found that exposure to ozone increases the allergen content of rye grass (*Lolium perenne*) anthers and pollen (Masuch *et al.*, 1997). In contrast to these findings, Albertine *et al.* (2014) found that increased exposure to a single pollutant (ozone) in an experimental setting resulted in a decrease in the amount of the allergen Phl p 5 expressed in timothy grass (*Phleum pretense* L.) pollen. However, the authors also found that exposure to elevated concentrations of CO_2 led to an increase in the amount of pollen produced by each flower, and that when timothy grass was grown under elevated concentrations of both ozone and CO_2, the increase in pollen production outweighed the decrease in allergen content. These results suggest that the net effect of increased ozone and CO_2 expected in the coming decades would be to increase exposure to timothy grass allergen (Albertine *et al.*, 2014).

Finally, there is evidence that air pollutants facilitate the release of allergens from pollen, increasing their presence in the atmosphere. For example, water-soluble compounds of DEP appear to enhance the release of allergen from pollen grains (Chehregani and Kouhkan, 2008). Similarly, timothy grass pollen exposed to NO_2 and ozone in a laboratory setting released more allergen-containing granules in a dose-dependent manner as the result of damage to the pollen exine (Motta *et al.*, 2006). However, the opposite effect has been observed for SO_2, which has been shown to have an inhibitory effect on allergen release (Behrendt and Becker, 2001). Additionally, while ozone may increase the release of allergen from the pollen grain, it has also been shown to inhibit pollen maturation in perennial rye grass (Schoene *et al.*, 2004). Thus, the interactions between air pollutants and pollen are competing and complex.

8.6 Conclusions and Implications for Action

Climate, air pollution, and aeroallergens interact in a variety of ways. At the level of atmospheric processes, air pollution levels can be elevated when inversions occur, when there is limited vertical and horizontal mixing, and when high temperatures favour ozone or secondary particle formation. Climate change is likely to affect all of these processes, leading to changing patterns of air pollution over space and time.

Due to interactions at the level of human dose and biological mechanisms, rising levels of CO_2, higher pollen levels and/or greater allergenicity, and exposures to traffic-related air pollution emissions such as DEP could lead to increasing health risks. This could be especially important in cities where CO_2 levels are higher than elsewhere (the 'CO_2 dome'), where traffic emissions are greatest, and where large numbers of vulnerable people live.

Because of the importance of cities as a locus of human exposure to both air pollution and airborne allergens, interventions to reduce greenhouse gases and air pollution emissions in cities via reduced motor vehicle traffic and/or emissions could lead to substantial benefits for human health as well as for climate mitigation. Air pollution control efforts targeting rapidly growing cities in the developing world will be especially important to address rising levels of air pollution and associated ill-health.

While bringing many benefits, urban tree planting programs undertaken in part to enhance localised cooling and air pollution capture have the potential to exacerbate allergic diseases via release of allergenic tree pollens. Care must be exercised to avoid increased pollen loads in these settings where the burden of existing disease is often highest. Selecting urban tree varieties that do not produce pollens of human health relevance is one potential strategy that may be effective in this regard.

References

Air Quality Expert Group (2007). *Air Quality and Climate Change: a UK Perspective*. London: Department for the Environment, Food and Rural Affairs.

Albertine, J. M., Manning, W. J., DaCosta, M., *et al.* (2014). Projected carbon dioxide to increase grass pollen and allergen exposure despite higher ozone levels. *PLoS One*, 9(11), e111712.

Andersen, Z. J., Bønnelykke, K., Hvidberg, M., *et al.* (2012). Long-term exposure to air pollution and asthma hospitalisations in older adults: a cohort study. *Thorax*, 67(1), 6–11.

Andersson, C., Engardt, M. (2010). European ozone in a future climate: importance of changes in dry deposition and isoprene emissions. *Journal of Geophysical Research: Atmospheres*, 115(D02), D02303.

Balmes, J. R., Fine, J. M., Sheppard, D. (1987). Symptomatic bronchoconstriction after short-term inhalation of sulfur dioxide. *American Review of Respiratory Disease*, 136(5), 1117–1121.

Bauer, M. A., Utell, M. J., Morrow, P. E., Speers, D. M., Gibb, F. R. (1986). Inhalation of 0.30 ppm nitrogen dioxide potentiates exercise-induced bronchospasm in asthmatics. *American Review of Respiratory Disease*, 134(5), 1203–1208.

Beck, I., Jochner, S., Gilles, S., *et al.* (2013). High environmental ozone levels lead to enhanced allergenicity of birch pollen. *PLoS One*, 8(11), e80147.

Behrendt, H., Becker, W.-M. (2001). Localization, release and bioavailability of pollen allergens: the influence of environmental factors. *Current Opinion in Immunology*, 13(6), 709–715.

Bell, M. L., McDermott, A., Zeger, S. L., Samet, J. M., Dominici, F. (2004). Ozone and short-term mortality in 95 US urban communities, 1987–2000. *Journal of the American Medical Association*, 292(19), 2372–2378.

Bell, M. L., Zanobetti, A., Dominici, F. (2014). Who is more affected by ozone pollution? A systematic review and meta-analysis. *American Journal of Epidemiology*, 180(1), 15–28.

Bentayeb, M., Simoni, M., Baiz, N., *et al.* (2012). Adverse respiratory effects of outdoor air pollution in the elderly. *The International Journal of Tuberculosis and Lung Disease*, 16(9), 1149–1161.

Bind, M.-A., Baccarelli, A., Zanobetti, A., *et al.* (2012). Air pollution and markers of coagulation, inflammation, and endothelial function: associations and epigene-environment interactions in an elderly cohort. *Epidemiology*, 23(2), 332–340.

Brunst, K. J., Leung, Y.-K., Ryan, P. H., *et al.* (2013). Forkhead box protein 3 (*FOXP3*) hypermethylation is associated with diesel exhaust exposure and risk for childhood asthma. *The Journal of Allergy and Clinical Immunology*, 131(2), 592–594.e3.

Burnett, R. T., Brook, J. R., Yung, W. T., Dales, R. E., Krewski, D. (1997). Association between ozone and hospitalization for respiratory diseases in 16 Canadian cities. *Environmental Research*, 72(1), 24–31.

Carvalho, A., Monteiro, A., Solman, S., Miranda, A. I., Borrego, C. (2010). Climate-driven changes in air quality over Europe by the end of the 21st century, with special reference to Portugal. *Environmental Science & Policy*, 13(6), 445–458.

Cecchi, L., D'amato, G., Ayres, J. G., *et al.* (2010). Projections of the effects of climate change on allergic asthma: the contribution of aerobiology. *Allergy*, 65(9), 1073–1081.

Chang, H. H., Zhou, J., Fuentes, M. (2010). Impact of climate change on ambient ozone level and mortality in southeastern United States. *International Journal of Environmental Research and Public Health*, 7(7), 2866–2880.

Chehregani, A., Kouhkan, F. (2008). Diesel exhaust particles and allergenicity of pollen grains of *Lilium martagon*. *Ecotoxicology and Environmental Safety*, 69(3), 568–573.

Chen, C.-H., Chan, C.-C., Chen, B.-Y., Cheng, T.-J., Guo, Y. L. (2015). Effects of particulate air pollution and ozone on lung function in non-asthmatic children. *Environmental Research*, 137, 40–48.

Chen, R., Cai, J., Meng, X., *et al.* (2014). Ozone and daily mortality rate in 21 cities of East Asia: how does season modify the association? *American Journal of Epidemiology*, 180(7), 729–736.

Chu, S., Zhang, H., Maher, C., *et al.* (2013). Prenatal and postnatal polycyclic aromatic hydrocarbon exposure, airway hyperreactivity, and beta-2 adrenergic receptor function in sensitized mouse offspring. *Journal of Toxicology*, 2013, 603581.

Cortegano, I., Civantos, E., Aceituno, E., *et al.* (2004). Cloning and expression of a major allergen from *Cupressus arizonica* pollen, Cup a 3, a PR-5 protein expressed under polluted environment. *Allergy*, 59(5), 485–490.

Dawson, J. P., Adams, P. J., Pandis, S. N. (2007a). Sensitivity of ozone to summertime climate in the eastern USA: a modeling case study. *Atmospheric Environment*, 41(7), 1494–1511.

Dawson, J. P., Adams, P. J., Pandis, S. N. (2007b). Sensitivity of $PM_{2.5}$ to climate in the Eastern US: a modeling case study. *Atmospheric Chemistry and Physics*, 7(16), 4295–4309.

Dawson, J. P., Racherla, P. N., Lynn, B. H., Adams, P. J., Pandis, S. N. (2008). Simulating present-day and future air quality as climate changes: model evaluation. *Atmospheric Environment*, 42(19), 4551–4566.

Delfino, R. J., Coate, B. D., Zeiger, R. S., *et al.* (1996). Daily asthma severity in relation to personal ozone exposure and outdoor fungal spores. *American Journal of Respiratory and Critical Care Medicine*, 154(3), 633–641.

Devalia, J. L., Rusznak, C., Wang, J., *et al.* (1996). Air pollutants and respiratory hypersensitivity. *Toxicology Letters*, 86(2–3), 169–176.

Devlin, R. B., McDonnell, W. F., Mann, R., *et al.* (1991). Exposure of humans to ambient levels of ozone for 6.6 hours causes cellular and biochemical changes in the lung. *American Journal of Respiratory Cell and Molecular Biology*, 4(1), 72–81.

Diaz-Sanchez, D., Garcia, M. P., Wang, M., Jyrala, M., Saxon, A. (1999). Nasal challenge with diesel exhaust particles can induce sensitization to a neoallergen in the human mucosa. *The Journal of Allergy and Clinical Immunology*, 104(6), 1183–1188.

Diaz-Sanchez, D., Tsien, A., Casillas, A., Dotson, A. R., Saxon, A. (1996). Enhanced nasal cytokine production in human beings after in vivo challenge with diesel exhaust particles. *The Journal of Allergy and Clinical Immunology*, 98(1), 114–123.

Diaz-Sanchez, D., Tsien, A., Fleming, J., Saxon, A. (1997). Combined diesel exhaust particulate and ragweed allergen challenge markedly enhances human in vivo nasal ragweed-specific IgE and skews cytokine production to a T helper cell 2-type pattern. *The Journal of Immunology*, 158(5), 2406–2413.

Driessen, M. N. B. M., Quanjer, Ph. H. (1991). Pollen deposition in intrathoracic airways. *European Respiratory Journal*, 4(3), 359–363.

Factor, P., Akhmedov, A. T., McDonald, J. D., *et al.* (2011). Polycyclic aromatic hydrocarbons impair function of β_2-adrenergic receptors in airway epithelial and smooth muscle cells. *American Journal of Respiratory Cell and Molecular Biology*, 45(5), 1045–1049.

Fang, Y., Mauzerall, D. L., Liu, J., Fiore, A. M., Horowitz, L. W. (2013a). Impacts of 21st century climate change on global air pollution-related premature mortality. *Climatic Change*, 121(2), 239–253.

Fang, Y., Naik, V., Horowitz, L. W., Mauzerall, D. L. (2013b). Air pollution and associated human mortality: the role of air pollutant emissions, climate change and methane concentration increases from the preindustrial period to present. *Atmospheric Chemistry and Physics*, 13(3), 1377–1394.

Fernvik, E., Peltre, G., Sénéchal, H., Vargaftig, B. B. (2002). Effects of birch pollen and traffic particulate matter on Th2 cytokines, immunoglobulin E levels and bronchial hyper-responsiveness in mice. *Clinical and Experimental Allergy*, 32(4), 602–611.

Forkel, R., Knoche, R. (2006). Regional climate change and its impact on photooxidant concentrations in southern Germany: simulations with a coupled regional climate-chemistry model. *Journal of Geophysical Research: Atmospheres*, 111(D12), D12302.

Fu, A., Leaderer, B. P., Gent, J. F., Leaderer, D., Zhu, Y. (2012). An environmental epigenetic study of *ADRB2* 5′-UTR methylation and childhood asthma severity. *Clinical and Experimental Allergy*, 42(11), 1575–1581.

Fujimaki, H., Nohara, O., Ichinose, T., Watanabe, N., Saito, S. (1994). IL-4 production in mediastinal lymph node cells in mice intratracheally instilled with diesel exhaust particulates and antigen. *Toxicology*, 92(1–3), 261–268.

Gao, Y., Fu, J. S., Drake, J. B., Lamarque, J.-F., Liu, Y. (2013). The impact of emission and climate change on ozone in the United States under representative concentration pathways (RCPs). *Atmospheric Chemistry and Physics*, 13(18), 9607–9621.

Ghiani, A., Aina, R., Asero, R., Bellotto, E., Citterio, S. (2012). Ragweed pollen collected along high-traffic roads shows a higher allergenicity than pollen sampled in vegetated areas. *Allergy*, 67(7), 887–894.

Gilliland, F. D., Li, Y.-F., Saxon, A., Diaz-Sanchez, D. (2004). Effect of glutathione-S-transferase M1 and P1 genotypes on xenobiotic enhancement of allergic responses: randomised, placebo-controlled crossover study. *The Lancet*, 363(9403), 119–125.

Gosling, S. N., McGregor, G. R., Lowe, J. A. (2012). The benefits of quantifying climate model uncertainty in climate change impacts assessment: an example with heat-related mortality change estimates. *Climatic Change*, 112(2), 217–231.

Gosling, S. N., McGregor, G. R., Páldy, A. (2007). Climate change and heat-related mortality in six cities Part 1: model construction and validation. *International Journal of Biometeorology*, 51(6), 525–540.

Granum, B., Gaarder, P. I., Løvik, M. (2001). IgE adjuvant effect caused by particles – immediated and delayed effects. *Toxicology*, 156(2–3), 149–159.

Gruzieva, O., Gehring, U., Aalberse, R., *et al.* (2014). Meta-analysis of air pollution exposure association with allergic sensitization in European birth cohorts. *The Journal of Allergy and Clinical Immunology*, 133(3), 767–776.

Halonen, J. I., Lanki, T., Yli-Tuomi, T., *et al.* (2009). Particulate air pollution and acute cardiorespiratory hospital admissions and mortality among the elderly. *Epidemiology*, 20(1), 143–153.

Hamada, K., Suzaki, Y., Leme, A., *et al.* (2007). Exposure of pregnant mice to an air pollutant aerosol increases asthma susceptibility in offspring. *Journal of Toxicology and Environmental Health, Part A*, 70(8), 688–695.

Hedegaard, G. B., Christensen, J. H., Brandt, J. (2013). The relative importance of impacts from climate change vs. emissions change on air pollution levels in the 21st century. *Atmospheric Chemistry and Physics*, 13(7), 3569–3585.

Heinrich, J., Hoelscher, B., Wjst, M., *et al.* (1999). Respiratory diseases and allergies in two polluted areas in East Germany. *Environmental Health Perspectives*, 107(1), 53–62.

Heo, Y., Saxon, A., Hankinson, O. (2001). Effect of diesel exhaust particles and their components on the allergen-specific IgE and IgG1 response in mice. *Toxicology*, 159(3), 143–158.

Hogrefe, C., Leung, R., Mickley, L., Hunt, S., Winner, D. (2005). Considering climate change in air quality management. *EM [Environmental Manager, Air & Waste Management Association]*, October, 19–23.

Hogrefe, C., Lynn, B., Civerolo, K., *et al.* (2004). Simulating changes in regional air pollution over the eastern United States due to changes in global and regional climate and emissions. *Journal of Geophysical Research: Atmospheres*, 109(D22), D22301.

Horstman, D. H., Folinsbee, L. J., Ives, P. J., Abdul-Salaam, S., McDonnell, W. F. (1990). Ozone concentration and pulmonary response relationships for 6.6-hour exposures with five hours of moderate exercise to 0.08, 0.10, and 0.12 ppm. *American Review of Respiratory Disease*, 142(5), 1158–1163.

Horton, D. E., Harshvardhan, Diffenbaugh, N. S. (2012). Response of air stagnation frequency to anthropogenically enhanced radiative forcing. *Environmental Research Letters*, 7(4), 044034.

Howlett, B. J., Knox, R. B. (1984). Allergic interactions. In: Linskens, H. F., Heslop-Harrison, J., eds. *Cellular Interactions. Encyclopedia of Plant Physiology, New Series, Volume 17*. Berlin: Springer-Verlag, pp. 655–673.

Huang, H.-C., Liang, X.-Z., Kunkel, K. E., Caughey, M., Williams, A. (2007). Seasonal simulation of tropospheric ozone over the midwestern and northeastern United States: an application of a coupled regional climate and air quality modeling system. *Journal of Applied Meteorology and Climatology*, 46(7), 945–960.

Inoue, K., Takano, H., Sakurai, M., *et al.* (2006). Pulmonary exposure to diesel exhaust particles enhances coagulatory disturbance with endothelial damage and systemic inflammation related to lung inflammation. *Experimental Biology and Medicine*, 231(10), 1626–1632.

Ishizaki, T., Koizumi, K., Ikemori, R., Ishiyama, Y., Kushibiki, E. (1987). Studies of prevalence of Japanese cedar pollinosis among the residents in a densely cultivated area. *Annals of Allergy*, 58(4), 265–270.

Jacob, D. J. (2005). *Interactions of Climate Change and Air Quality: Research Priorities and New Direction*. Report from a Workshop, 26–27 April 2005, Washington, DC. Program on Technology Innovation, 1012169. Palo Alto, CA: Electric Power Research Institute.

Jacob, D. J., Winner, D. A. (2009). Effect of climate change on air quality. *Atmospheric Environment*, 43(1), 51–63.

Jacobson, M. Z. (2008). On the causal link between carbon dioxide and air pollution mortality. *Geophysical Research Letters*, 35(3), L03809.

Jerrett, M., Shankardass, K., Berhane, K., *et al.* (2008). Traffic-related air pollution and asthma onset in children: a prospective cohort study with individual exposure measurement. *Environmental Health Perspectives*, 116(10), 1433–1438.

Jhun, I., Fann, N., Zanobetti, A., Hubbell, B. (2014). Effect modification of ozone-related mortality risks by temperature in 97 US cities. *Environment International*, 73, 128–134.

Jung, K. H., Patel, M. M., Moors, K., *et al.* (2010). Effects of heating season on residential indoor and outdoor polycyclic aromatic hydrocarbons, black carbon, and particulate matter in an urban birth cohort. *Atmospheric Environment*, 44(36), 4545–4552.

Jung, K. H., Yan, B., Moors, K., *et al.* (2012). Repeated exposure to polycyclic aromatic hydrocarbons and asthma: effect of seroatopy. *Annals of Allergy, Asthma & Immunology*, 109(4), 249–254.

Kadkhoda, K., Pourfathollah, A. A., Pourpak, Z., Kazemnejad, A. (2005). The cumulative activity of benzo(*a*)pyrene on systemic immune responses with mite allergen extract after intranasal instillation and ex vivo response to ovalbumin in mice. *Toxicology Letters*, 157(1), 31–39.

Kahle, J. J., Neas, L. M., Devlin, R. B., *et al.* (2015). Interaction effects of temperature and ozone on lung function and markers of systemic inflammation, coagulation, and fibrinolysis: a crossover study of healthy young volunteers. *Environmental Health Perspectives*, 123(4), 310–316.

Kan, H., Chen, R., Tong, S. (2012). Ambient air pollution, climate change, and population health in China. *Environment International*, 42, 10–19.

Kanoh, T., Suzuki, T., Ishimori, M., *et al.* (1996). Adjuvant activities of pyrene, anthracene, fluoranthene and benzo(a)pyrene in production of anti-IgE antibody to Japanese cedar pollen allergen in mice. *Journal of Clinical & Laboratory Immunology*, 48(4), 133–147.

Katragkou, E., Zanis, P., Kioutsioukis, I., *et al.* (2011). Future climate change impacts on summer surface ozone from regional climate-air quality simulations over Europe. *Journal of Geophysical Research: Atmospheres*, 116(D22), D22307.

Kinney, P. L. (2008). Climate change, air quality, and human health. *American Journal of Preventive Medicine*, 35(5), 459–467.

Kinney, P. L., Nilsen, D. M., Lippmann, M., *et al.* (1996). Biomarkers of lung inflammation in recreational joggers exposed to ozone. *American Journal of Respiratory and Critical Care Medicine*, 154(5), 1430–1435.

Kinney, P. L., Özkaynak, H. (1991). Associations of daily mortality and air pollution in Los Angeles County. *Environmental Research*, 54(2), 99–120.

Kinney, P. L., Ware, J. H., Spengler, J. D., *et al.* (1989). Short-term pulmonary function change in association with ozone levels. *American Review of Respiratory Disease*, 139(1), 56–61.

Knowlton, K., Hogrefe, C., Lynn, B., *et al.* (2008). Impacts of heat and ozone on mortality risk in the New York City metropolitan region under a changing climate. In: Thomson, M. C., Garcia-Herrera, R., Beniston, M., eds. *Seasonal Forecasts, Climatic Change and Human Health*. The Netherlands: Springer, pp. 143–160.

Knowlton, K., Rosenthal, J. E., Hogrefe, C., *et al.* (2004). Assessing ozone-related health impacts under a changing climate. *Environmental Health Perspectives*, 112(15), 1557–1563.

Knox, R. B., Suphioglu, C., Taylor, P., *et al.* (1997). Major grass pollen allergen Lol p 1 binds to diesel exhaust particles: implications for asthma and air pollution. *Clinical and Experimental Allergy*, 27(3), 246–251.

Kobayashi, T. (2000). Exposure to diesel exhaust aggravates nasal allergic reaction in guinea pigs. *American Journal of Respiratory and Critical Care Medicine*, 162(2), 352–356.

Künzli, N., Perez, L., Rapp, R. (2010). *Air Quality and Health*. Lausanne, Switzerland: European Respiratory Society.

Lam, Y. F., Fu, J. S., Wu, S., Mickley, L. J. (2011). Impacts of future climate change and effects of biogenic emissions on surface ozone and particulate matter concentrations in the United States. *Atmospheric Chemistry and Physics*, 11(10), 4789–4806.

Langner, J., Engardt, M., Baklanov, A., *et al.* (2012). A multi-model study of impacts of climate change on surface ozone in Europe. *Atmospheric Chemistry and Physics*, 12(21), 10423–10440.

Leung, L. R., Gustafson Jr, W. I. (2005). Potential regional climate change and implications to U.S. air quality. *Geophysical Research Letters*, 32(16), L16711.

Li, T., Yan, M., Ma, W., *et al.* (2015). Short-term effects of multiple ozone metrics on daily mortality in a megacity of China. *Environmental Science and Pollution Research*, 22(11), 8738–8746.

Liang, X.-Z., Pan, J., Zhu, J., *et al.* (2006). Regional climate model downscaling of the U.S. summer climate and future change. *Journal of Geophysical Research: Atmospheres*, 111(D10), D10108.

Liao, K.-J., Tagaris, E., Russell, A. G., *et al.* (2010). Cost analysis of impacts of climate change on regional air quality. *Journal of the Air & Waste Management Association*, 60(2), 195–203.

Likhvar, V. N., Pascal, M., Markakis, K., *et al.* (2015). A multi-scale health impact assessment of air pollution over the 21st century. *Science of the Total Environment*, 514, 439–449.

Lin, S., Bell, E. M., Liu, W., *et al.* (2008). Ambient ozone concentration and hospital admissions due to childhood respiratory diseases in New York State, 1991–2001. *Environmental Research*, 108(1), 42–47.

Liu, J., Ballaney, M., Al-alem, U., *et al.* (2008). Combined inhaled diesel exhaust particles and allergen exposure alter methylation of T helper genes and IgE production *in vivo*. *Toxicological Sciences*, 102(1), 76–81.

Liu, J., Zhang, L., Winterroth, L. C., *et al.* (2013). Epigenetically mediated pathogenic effects of phenanthrene on regulatory T cells. *Journal of Toxicology*, 2013, 967029.

Løvik, M., Høgseth, A.-K., Gaarder, P. I., Hagemann, R., Eide, I. (1997). Diesel exhaust particles and carbon black have adjuvant activity on the local lymph node response and systemic IgE production to ovalbumin. *Toxicology*, 121(2), 165–178.

Maejima, K., Tamura, K., Taniguchi, Y., Nagase, S., Tanaka, H. (1997). Comparison of the effects of various fine particles on IgE antibody production in mice inhaling Japanese cedar pollen allergens. *Journal of Toxicology and Environmental Health*, 52(3), 231–248.

Mahmud, A., Hixson, M., Kleeman, M. J. (2012). Quantifying population exposure to airborne particulate matter during extreme events in California due to climate change. *Atmospheric Chemistry and Physics*, 12(16), 7453–7463.

Masuch, G., Franz, J.-Th., Schoene, K., Müsken, H., Bergmann, K.-Ch. (1997). Ozone increases group 5 allergen content of *Lolium perenne*. *Allergy*, 52(8), 874–875.

Matsumoto, A., Hiramatsu, K., Li, Y., *et al.* (2006). Repeated exposure to low-dose diesel exhaust after allergen challenge exaggerates asthmatic responses in mice. *Clinical Immunology*, 121(2), 227–235.

McClellan, R. O. (1987). Health effects of exposure to diesel exhaust particles. *Annual Review of Pharmacology and Toxicology*, 27, 279–300.

Mickley, L. J., Jacob, D. J., Field, B. D., Rind, D. (2004). Effects of future climate change on regional air pollution episodes in the United States. *Geophysical Research Letters*, 31(24), L24103.

Miller, R. L., Garfinkel, R., Horton, M., *et al.* (2004). Polycyclic aromatic hydrocarbons, environmental tobacco smoke, and respiratory symptoms in an inner-city birth cohort. *Chest*, 126(4), 1071–1078.

Miller, R. L., Garfinkel, R., Lendor, C., *et al.* (2010). Polycyclic aromatic hydrocarbon metabolite levels and pediatric allergy and asthma in an inner-city cohort. *Pediatric Allergy and Immunology*, 21(2), 260–267.

Miller, R. L., Peden, D. B. (2014). Environmental effects on immune responses in patients with atopy and asthma. *The Journal of Allergy and Clinical Immunology*, 134(5), 1001–1008.

Morgenstern, V., Zutavern, A., Cyrys, J., *et al.* (2008). Atopic diseases, allergic sensitization, and exposure to traffic-related air pollution in children. *American Journal of Respiratory and Critical Care Medicine*, 177(12), 1331–1337.

Motta, A. C., Marliere, M., Peltre, G., Sterenberg, P. A., Lacroix, G. (2006). Traffic-related air pollutants induce the release of allergen-containing cytoplasmic granules from grass pollen. *International Archives of Allergy and Immunology*, 139(4), 294–298.

Muranaka, M., Suzuki, S., Koizumi, K., *et al.* (1986). Adjuvant activity of diesel-exhaust particulates for the production of IgE antibody in mice. *The Journal of Allergy and Clinical Immunology*, 77(4), 616–623.

Murazaki, K., Hess, P. (2006). How does climate change contribute to surface ozone change over the United States? *Journal of Geophysical Research: Atmospheres*, 111(D05), D05301.

Nadeau, K., McDonald-Hyman, C., Noth, E. M., *et al.* (2010). Ambient air pollution impairs regulatory T-cell function in asthma. *The Journal of Allergy and Clinical Immunology*, 126(4), 845–852.e10.

Namork, E., Johansen, B. V., Løvik, M. (2006). Detection of allergens adsorbed to ambient air particles collected in four European cities. *Toxicology Letters*, 165(1), 71–78.

Nolte, C. G., Gilliland, A. B., Hogrefe, C., Mickley, L. J. (2008). Linking global to regional models to assess future climate impacts on surface ozone levels in the United States. *Journal of Geophysical Research: Atmospheres*, 113(D14), D14307.

Nygaard, U. C., Ormstad, H., Aase, A., Løvik, M. (2005). The IgE adjuvant effect of particles: characterisation of the primary cellular response in the draining lymph node. *Toxicology*, 206(2), 181–193.

Nygaard, U. C., Samuelsen, M., Aase, A., Løvik, M. (2004). The capacity of particles to increase allergic sensitization is predicted by particle number and surface area, not by particle mass. *Toxicological Sciences*, 82(2), 515–524.

Ormstad, H. (2000). Suspended particulate matter in indoor air: adjuvants and allergen carriers. *Toxicology*, 152(1–3), 53–68.

Ormstad, H., Johansen, B. V., Gaarder, P. I. (1998). Airborne house dust particles and diesel exhaust particles as allergen carriers. *Clinical and Experimental Allergy*, 28(6), 702–708.

Orru, H., Andersson, C., Ebi, K. L., *et al.* (2013). Impact of climate change on ozone-related mortality and morbidity in Europe. *European Respiratory Journal*, 41(2), 285–294.

Patel, M. M., Quinn, J. W., Jung, K. H., *et al.* (2011). Traffic density and stationary sources of air pollution associated with wheeze, asthma, and immunoglobulin E from birth to age 5 years among New York City children. *Environmental Research*, 111(8), 1222–1229.

Perzanowski, M. S., Chew, G. L., Divjan, A., *et al.* (2013). Early-life cockroach allergen and polycyclic aromatic hydrocarbon exposures predict cockroach sensitization among inner-city children. *The Journal of Allergy and Clinical Immunology*, 131(3), 886–893.e6.

Petersen, A., Schramm, G., Schlaak, M., Becker, W.-M. (1998). Post-translational modifications influence IgE reactivity to the major allergen Phl p 1 of timothy grass pollen. *Clinical and Experimental Allergy*, 28(3), 315–321.

Pope III, C. A., Dockery, D. W. (2006). Health effects of fine particulate air pollution: lines that connect. *Journal of the Air & Waste Management Association*, 56(6), 709–742.

Post, E. S., Grambsch, A., Weaver, C., *et al.* (2012). Variation in estimated ozone-related health impacts of climate change due to modeling choices and assumptions. *Environmental Health Perspectives*, 120(11), 1559–1564.

Prestera, T., Talalay, P. (1995). Electrophile and antioxidant regulation of enzymes that detoxify carcinogens. *Proceedings of the National Academy of Sciences of the United States of America*, 92(19), 8965–8969.

Racherla, P. N., Adams, P. J. (2006). Sensitivity of global tropospheric ozone and fine particulate matter concentrations to climate change. *Journal of Geophysical Research: Atmospheres*, 111(D24), D24103.

Racherla, P. N., Adams, P. J. (2008). The response of surface ozone to climate change over the eastern United States. *Atmospheric Chemistry and Physics*, 8(4), 871–885.

Racherla, P. N., Adams, P. J. (2009). U.S. ozone air quality under changing climate and anthropogenic emissions. *Environmental Science & Technology*, 43(3), 571–577.

Rogerieux, F., Godfrin, D., Sénéchal, H., *et al.* (2007). Modifications of *Phleum pratense* grass pollen allergens following artificial exposure to gaseous air pollutants (O_3, NO_2, SO_2). *International Archives of Allergy and Immunology*, 143(2), 127–134.

Rosa, M. J., Jung, K. H., Perzanowski, M. S., *et al.* (2011). Prenatal exposure to polycyclic aromatic hydrocarbons, environmental tobacco smoke and asthma. *Respiratory Medicine*, 105(6), 869–876.

Samuelsen, M., Nygaard, U. C., Løvik, M. (2008). Allergy adjuvant effect of particles from wood smoke and road traffic. *Toxicology*, 246(2–3), 124–131.

Schoene, K., Franz, J.-Th., Masuch, G. (2004). The effect of ozone on pollen development in *Lolium perenne* L. *Environmental Pollution*, 131(3), 347–354.

Sheffield, P. E., Knowlton, K., Carr, J. L., Kinney, P. L. (2011). Modeling of regional climate change effects on ground-level ozone and childhood asthma. *American Journal of Preventive Medicine*, 41(3), 251–257.

Silverman, R. A., Ito, K. (2010). Age-related association of fine particles and ozone with severe acute asthma in New York City. *The Journal of Allergy and Clinical Immunology*, 125(2), 367–373.

Steerenberg, P. A., Dormans, J. A. M. A., van Doorn, C. C. M., *et al.* (1999). A pollen model in the rat for testing adjuvant activity of air pollution components. *Inhalation Toxicology*, 11(12), 1109–1122.

Steiner, A. L., Tonse, S., Cohen, R. C., Goldstein, A. H., Harley, R. A. (2006). Influence of future climate and emissions on regional air quality in California. *Journal of Geophysical Research: Atmospheres*, 111(D18), D18303.

Steinsvik, T. E., Ormstad, H., Gaarder, P. I., *et al.* (1998). Human IgE production in hu-PBL-SCID mice injected with birch pollen and diesel exhaust particles. *Toxicology*, 128(3), 219–230.

Stevenson, D. S., Dentener, F. J., Schultz, M. G., *et al.* (2006). Multimodel ensemble simulations of present-day and near-future tropospheric ozone. *Journal of Geophysical Research: Atmospheres*, 111(D08), D08301.

Suárez-Cervera, M., Castells, T., Vega-Maray, A., *et al.* (2008). Effects of air pollution on Cup a 3 allergen in *Cupressus arizonica* pollen grains. *Annals of Allergy, Asthma & Immunology*, 101(1), 57–66.

Sujaritpong, S., Dear, K., Cope, M., Walsh, S., Kjellstrom, T. (2014). Quantifying the health impacts of air pollution under a changing climate – a review of approaches and methodology. *International Journal of Biometeorology*, 58(2), 149–160.

Suzuki, T., Kanoh, T., Ishimori, M., Ikeda, S., Ohkuni, H. (1996). Adjuvant activity of diesel exhaust particulates (DEP) in production of anti-IgE and anti-IgG1 antibodies to mite allergen in mice. *Journal of Clinical and Laboratory Immunology*, 48(5), 187–199.

Tagaris, E., Liao, K.-J., Delucia, A. J., *et al.*, (2009). Potential impact of climate change on air pollution-related human health effects. *Environmental Science & Technology*, 43(13), 4979–4988.

Tagaris, E., Liao, K.-J., DeLucia, A. J., *et al.* (2010). Sensitivity of air pollution-induced premature mortality to precursor emissions under the influence of climate change. *International Journal of Environmental Research and Public Health*, 7(5), 2222–2237.

Tagaris, E., Liao, K.-J., Manomaiphiboon, K., *et al.* (2008). The role of climate and emission changes in future air quality over southern Canada and northern Mexico. *Atmospheric Chemistry and Physics*, 8(14), 3973–3983.

Tagaris, E., Manomaiphiboon, K., Liao, K.-J., *et al.* (2007). Impacts of global climate change and emissions on regional ozone and fine particulate matter concentrations over the United States. *Journal of Geophysical Research: Atmospheres*, 112(D14), D14312.

Tai, A. P. K., Mickley, L. J., Jacob, D. J. (2012). Impact of 2000–2050 climate change on fine particulate matter ($PM_{2.5}$) air quality inferred from a multi-model analysis of meteorological modes. *Atmospheric Chemistry and Physics*, 12(23), 11329–11337.

Takafuji, S., Suzuki, S., Koizumi, K., *et al.* (1987). Diesel-exhaust particulates inoculated by the intranasal route have an adjuvant activity for IgE production in mice. *The Journal of Allergy and Clinical Immunology*, 79(4), 639–645.

Takahashi, G., Tanaka, H., Wakahara, K., *et al.* (2010). Effect of diesel exhaust particles on house dust mite-induced airway eosinophilic inflammation and remodeling in mice. *Journal of Pharmacological Sciences*, 112(2), 192–202.

Takenaka, H., Zhang, K., Diaz-Sanchez, D., Tsien, A., Saxon, A. (1995). Enhanced human IgE production results from exposure to the aromatic hydrocarbons from diesel exhaust: direct effects on B-cell IgE production. *The Journal of Allergy and Clinical Immunology*, 95(1), 103–115.

Tao, Z., Williams, A., Huang, H.-C., Caughey, M., Liang, X.-Z. (2007). Sensitivity of U.S. surface ozone to future emissions and climate changes. *Geophysical Research Letters*, 34(L08), L08811.

The Royal Society (2008). *Ground-Level Ozone in the 21st Century: Future Trends, Impacts and Policy Implications. Science Policy Report 15/08*. London: The Royal Society.

Weaver, C. P., Liang, X.-Z., Zhu, J., *et al.* (2009). A preliminary synthesis of modeled climate change impacts on U.S. regional ozone concentrations. *Bulletin of the American Meteorological Society*, 90(12), 1843–1863.

Westerholm, R. N., Almén, J., Li, H., *et al.* (1991). Chemical and biological characterization of particulate-, semivolatile-, and gas-phase-associated compounds in diluted heavy-duty diesel exhausts: a comparison of three different semivolatile-phase samplers. *Environmental Science & Technology*, 25(2), 332–338.

Winquist, A., Klein, M., Tolbert, P., *et al.* (2012). Comparison of emergency department and hospital admissions data for air pollution time-series studies. *Environmental Health*, 11, 70.

World Health Organization (WHO) (2013). *Review of Evidence on Health Aspects of Air Pollution – REVIHAAP Project: Technical Report*. Copenhagen: World Health Organization Regional Office for Europe.

Wu, S., Mickley, L. J., Leibensperger, E. M., *et al.* (2008). Effects of 2000–2050 global change on ozone air quality in the United States. *Journal of Geophysical Research: Atmospheres*, 113(D06), D06302.

Yoshida, T., Yoshioka, Y., Fujimura, M., *et al.* (2010). Potential adjuvant effect of intranasal urban aerosols in mice through induction of dendritic cell maturation. *Toxicology Letters*, 199(3), 383–388.

Zeng, G., Pyle, J. A. (2003). Changes in tropospheric ozone between 2000 and 2100 modeled in a chemistry-climate model. *Geophysical Research Letters*, 30(7), 1392.

9

Impacts of Climate Change on Allergic Diseases

CONSTANCE H. KATELARIS[1,2]
[1]School of Medicine
Western Sydney University
[2]Immunology and Allergy
Department of Medicine
Campbelltown Hospital

9.1 Introduction

Allergic diseases result from complex interactions between genes and environment. Environmental factors that are known to initiate and/or exacerbate allergic disorders, particularly allergic respiratory diseases, include exposure to pollen, fungi, and pollutants (Kim and Bernstein, 2009; Tham *et al.*, 2014; Traidl-Hoffmann *et al.*, 2003). Thus, pollen and fungal spore production and distribution have a major impact on incidence, severity, and pattern of allergic respiratory symptoms. Previous chapters have detailed the many changes that have occurred, and are likely to occur, as a consequence of our changing climate; these changes impact on all living organisms, changing their physiology and distribution.

Climate change will influence aeroallergen sensitivity in many ways; evidence exists for an impact on expression of allergenic proteins within pollen grains; some plants at least will respond to increased carbon dioxide (CO_2) levels with increased pollen production (Singer *et al.*, 2005; Ziska and Caulfield, 2000).

Expected changes in rainfall patterns will result in greater rainfall in some regions with consequent increase in vegetation, while in other regions, decreased precipitation will result in loss of vegetation and decrease in aeroallergen exposure. Indeed, changes along these lines are already evident and changes in aeroallergen distribution have already been reported (Settele *et al.*, 2014; Storkey *et al.*, 2014).

An increase in extreme weather events such as tropical cyclones (hurricanes), thunderstorms, and dust storms are projected based on various climate change scenarios. There is existing evidence that these events have significant impact on the initiation and exacerbation of allergic respiratory disease (D'Amato and Cecchi, 2008).

Various atmospheric pollutants are increasing and, without mitigation, will continue to do so (Smith *et al.*, 2014). This is the result of increasing urbanisation and industrialisation on a vast scale as is occurring in China and other emerging nations. Continued growth in vehicular traffic volumes results in increase in particulate pollutants, especially diesel exhaust particles (DEPs). Pollutants have complex and diverse influences on allergic respiratory diseases. Evidence exists that exposure to various particulate and gaseous pollutants increases inflammatory processes in the human airway leading to impairment of lung development in young children and initiation of asthma. Asthma exacerbations and increased numbers of presentations to emergency departments (ED) are also documented in relation to increases in pollutant levels. A number of reviews have explored these many issues in depth (Beggs, 2004; Cecchi *et al.*, 2010; Reid and Gamble, 2009; Shea *et al.*, 2008; Weber, 2012).

The purpose of this chapter is to establish the importance of aeroallergen exposure in causation and triggering of allergic diseases such as asthma, allergic rhinoconjunctivitis, and atopic dermatitis (AD) and to explain how anticipated changes in climate may impact the prevalence, severity, and distribution of these conditions.

9.2 Importance/Burden of Allergic Diseases Globally

Allergic diseases are now recognised as one of the most prevalent and important non-communicable diseases worldwide (Ozdoganoglu and Songu, 2012). They are common conditions that have their peak prevalence in children and young adults and as such have a very significant impact on quality of life as well as create a very substantial economic burden at both individual and societal levels (Crystal-Peters *et al.*, 2000; Szeinbach *et al.*, 2007).

Allergic rhinitis (AR) affects between 10% and 30% of adults and as much as 40% of children (Masoli *et al.*, 2004). Epidemiologic studies show that the prevalence of AR continues to increase worldwide (Riedl and Diaz-Sanchez, 2005).

The World Health Organization (WHO) has estimated that 400 million people in the world suffer from AR, and 300 million from asthma. AR and asthma are linked conditions; approximately 40% of people with AR have asthma and over 80% of people with asthma have AR. AR is associated with, or is a risk factor for, other co-morbidities such as sinusitis, otitis media with effusion, nasal polyps, and sleep apnoea, further increasing the burden of this condition (Borish, 2003; Guerra *et al.*, 2002; Nathan, 2007).

Allergic conjunctivitis (AC) is most commonly seen in those with AR but may occur without nasal allergy. There are a number of forms of AC, the commonest being seasonal and perennial AC, which while not sight-threatening have significant impacts on quality of life. The more complex disorders are associated with chronic inflammation and the possibility of severe, sight-threatening injury to the eye (Bielory and Friedlaender, 2008; Katelaris, 2003).

Asthma is a serious public health problem throughout the world, affecting people of all ages. The Global Initiative for Asthma report (Akdis and Agache, 2013) summarises the WHO estimates that globally 300 million individuals have asthma and that a quarter of a million people die of this condition annually.

Three large international surveys on adult asthma have been conducted: European Community Respiratory Health Survey (ECRHS; 1991–1994; Burney *et al.*, 1996), World Health Survey (2002–2003; Sembajwe *et al.*, 2010), and Global Allergy and Asthma Network of Excellence (GA^{2}LEN; 2008–2009; Jarvis *et al.*, 2012). These studies differ in the manner in which data was collected, so it is difficult to compare data across the surveys. Each survey suggests substantial geographical variation in adult asthma prevalence between countries. Two of the studies, ECRHS and the GA^{2}LEN survey, provide some evidence of cohort-related increases in adult asthma.

The best information on childhood prevalence of asthma throughout the world was obtained by the International Study of Asthma and Allergies in Childhood (Asher *et al.*, 2006). Questionnaires were completed in 1994 and 1995 by 463,801 children aged 13–14 years from fifty-six countries, and by parents of 257,800 children aged 6–7 years from thirty-eight countries. In the 13–14-year-old age group, the indicated prevalence varied more than fifteen-fold between countries, ranging from 2.1% to 4.4% in Albania, China, Georgia, Indonesia, and Romania to 29.1% to 32.2% in Australia, New Zealand, Republic of Ireland, and the United Kingdom. Other countries with low prevalence were mostly in Asia, northern Africa, eastern Europe, and the eastern Mediterranean regions, and others with high prevalence were in southeast Asia, North America, and Latin America. Trends for prevalence in the 6–7-year-olds were similar to those in the older children with prevalence of wheezing varying from 4.1% to 32.1% (Asher *et al.*, 2006).

The economic burden of asthma worldwide is enormous and must be considered on a number of levels (Masoli *et al.*, 2004). The direct costs of disease comprise the health-care expenditure associated with hospitalisations, emergency visits, physician visits, diagnostic tests, and medical treatment of which the most important cost components are hospital admissions and asthma medication. Indirect costs include the impact on employment, loss of work productivity, and other social costs. The number of disability-adjusted life years (DALYs) lost due to asthma worldwide has been estimated to be currently about 15 million per year. Worldwide, asthma accounts for around 1% of all DALYs lost, which reflects the high prevalence and severity of asthma. In the United States, the estimated annual costs per patient with asthma is US$ 1,907 and the total national medical expenditure is US$ 18 billion. The total costs of asthma in Europe are estimated at approximately € 17.7 billion per annum (Akdis and Agache, 2013).

Australian, US, and Canadian studies have found that direct costs account for the greatest part of the total costs. However, the US TENOR study focussing on severe and difficult-to-treat asthma demonstrated higher indirect than direct costs (Chipps *et al.*, 2012). Also, several European studies among which is a large German study demonstrated that up to 75% of the total costs of asthma could be attributed to indirect costs (Akdis and Agache, 2013; Bahadori *et al.*, 2009).

9.3 Relationship of Aeroallergens to Allergic Respiratory Diseases

Observed and projected changes in climate will have profound effects on aeroallergen distribution and composition through diverse mechanisms.

9.3.1 Pollen, Allergic Sensitisation, and Respiratory Disease

Current projections suggest that overall there is an increase in pollen exposure as a result of climate change (Reid and Gamble, 2009; Shea *et al.*, 2008; Chapters 2–6, and 8). This may lead to an increase in allergic sensitisation and clinical expression of related diseases. Clinical evidence of the importance of pollen exposure in causation of allergic respiratory diseases comes from a variety of approaches, ranging from exploring the patterns of sensitisation to various pollen allergens, to examining correlations between the pollen count and asthma exacerbations, usually measured by hospital attendances or admissions (Erbas *et al.*, 2012). The latter is an important parameter as it carries significant economic implications.

9.3.1.1 Sensitisation

Development of pollen sensitisation depends on numerous factors including genetic susceptibility; level and length of exposure, i.e., duration of the pollen season in a given region; type of pollen and its allergenicity; and possibly also the relationship of early exposure to time of birth (Harley *et al.*, 2009).

There is little doubt that climatic variations will impact on most of these variables in a significant manner, and indeed there have already been observed changes to the pollen season in some locations, and this may be expected to impact on rates and degrees of sensitisation (Chapter 6; Ziska *et al.*, 2011).

In order to study these trends and changes, access to detailed and long-term aerobiological surveys is necessary, but unfortunately, there are relatively few such data sets in existence. However, there are some long-term phenological records in many European countries, and these demonstrate measurable changes occurring in recent decades. Flowering is particularly sensitive to temperature over the preceding month. Fitter and Fitter (2002) have demonstrated a 4.5-day advancement in the average first flowering date for nearly 400 British plant species for the last decade of the twentieth century (1991–2000) compared to the previous four decades (1954–1990). Menzel (2000) has examined a Europe-wide network, the European phenological database, and reports that the average annual growing season has lengthened by approximately 11 days since the 1960s.

Ariano *et al.* (2010) had a unique opportunity to study variations in pollen levels and allergic sensitisation in western Liguria, Italy, because of the existence of almost three decades of pollen-monitoring and meteorological variable data and skin prick test and clinical data from residents in the region. They describe a progressive increase in the duration of the pollen season for *Parietaria* (+85 days), olive (+18 days), and cypress (+18 days; Ariano *et al.*, 2010). All pollen monitored except for grasses showed an increase in total counts. They report an increase in the percentages of patients sensitised to pollen over these years, whereas sensitisation rates to house dust mite (HDM) remained stable.

Wheezing in young children is usually attributed to a viral aetiology, and sensitisation to outdoor allergens in this young age group has not been considered important. Ogershok *et al.* (2007) studied allergen sensitisation by performing skin prick tests with a battery of aeroallergens relevant to the particular population under study. Patients presenting with a history of asthma aged 0–10 years were studied. Of 687 children enrolled, 98 were under 3 years. The percentage of positive skin prick tests to outdoor allergens increased with increasing age: no infant under 1 year had positive tests to outdoor allergens; 29% aged 1–2 years reacted to a pollen; 40% of 2-year-olds and 49% of 3-year-olds reacted to one or more pollen allergens; and by 10 years, 70% of all children reacted to outdoor allergens, the same percentage that reacted to indoor allergens. In this population, the most common allergens were box elder, short ragweed, and June grass pollen (Ogershok *et al.*, 2007).

9.3.1.2 Seasonal Allergic Rhinoconjunctivitis

The relationship between seasonal rhinitis and conjunctivitis and pollen sensitisation and exposure is well established and demonstrated in many different locations with a range of different pollen species (Burr *et al.*, 2003; Charpin *et al.*, 1993). Relating pollen counts and severity of symptoms, and the amelioration of those symptoms with various medications, is the usual study format for pharmacologic agents developed for the treatment of seasonal allergic rhinoconjunctivitis (SARC). Tracing the onset and development of SARC in young children and adolescents is less well studied. In a prospective birth cohort, the Multicenter Allergy Study, Kulig *et al.* (2000) studied the development of SARC defined by a combination of sensitisation patterns and exposure history. They found that the occurrence of SARC increased progressively with increasing age from <2% in early childhood to 15% by 7 years.

9.3.1.3 Asthma Presentations and Pollen Counts

Asthma exacerbations are one of the most costly aspects of asthma management overall, as they often result in ED visits or hospital admissions. Thus, an important consideration regarding the impact of pollen levels on respiratory health outcomes is the relationship of hospitalisations to pollen levels. With anticipated changes to aeroallergen levels, possible longer growing seasons (Chapter 6) and other climatic changes that may result in a heavier aeroallergen burden (Chapters 2–8; Takaro *et al.*, 2013), more effort is being concentrated on understanding the effects of pollen on allergic respiratory disorders such as rhinitis and asthma.

While the pattern of aeroallergen sensitisation in those with asthma varies with economic status, ethnicity, and geographic location (Rastogi *et al.*, 2006), the majority of people with mild to moderately severe asthma show evidence of both indoor and outdoor aeroallergen sensitisation, and this correlates with a number of asthma characteristics (Craig, 2010). There are many published papers describing analyses of epidemiological data investigating the link between pollen levels and certain asthma-related outcomes. These investigations have yielded conflicting results (Carlsen *et al.*, 1984; Dales *et al.*, 2004; Darrow *et al.*, 2012; Rossi *et al.*, 1993), and the topic has been reviewed recently by Schmier and Ebi (2009). They highlight the heterogeneity of approaches, methodologies, and end points used in the numerous studies investigating this association. A brief summary of several investigations finding an association is given below.

In a large study in Madrid, Spain, Tobías *et al.* (2004) used Poisson regression with generalised additive models, controlling for trend and seasonality, meteorological variables, acute respiratory infections, and air pollutants, to examine risks for daily number of hospitalisations for asthma in 2,400 children. They found pollen levels were positively associated with hospitalisation even when adjustments were made for weather factors and air pollutant levels. Of the four pollen types examined (*Olea*, *Plantago*, Poaceae, and Urticaceae), the strongest association with asthma ED visits was with Poaceae.

In a study in Montreal, Canada, Breton *et al.* (2006) demonstrated a significant correlation between ragweed pollen levels and numbers of medical consultations for AR. Using time series analysis and adjusting for multiple possible confounders such as weather parameters and air pollutants, Héguy *et al.* (2008) found positive associations between ED visits for asthma exacerbations and pollen levels 3 days after exposure in Montreal, Canada.

Several studies have been conducted in the United States. Babin *et al.* (2007), studying asthma-related ED visits in children, found associations between tree and weed pollen levels and hospitalisations in Washington DC. Lierl and Hornung (2003), using multiple regression analysis, showed that the most significant predictor of ED visit for asthma in children living in Cincinnati was the 3-day lagged log of pollen count. Zhong *et al.* (2006) also demonstrated that lagged ragweed and tree pollen counts correlated best with ED visits in children in the Cincinnati area.

In Madrid, Spain, Galán *et al.* (2010) performed a case–control study of any individuals presenting to hospital with an asthma exacerbation, comparing skin prick test sensitisation to various pollen allergens with another group of people with asthma that did not present during the spring season. They found that sensitisation to grass pollen in particular was associated with asthma presentations during the peak in the spring.

In Melbourne, Australia, peaks in ED presentation for asthma exacerbation have been observed during spring months. Erbas *et al.* (2012), using a time series analysis, explored the relationship between grass pollen counts and asthma presentations to hospital, finding a relationship between increased risk of asthma exacerbation and level of grass pollen.

9.3.2 Fungi and Allergic Respiratory Disease

Allergic sensitisation to airborne fungi has been appreciated for decades, but the relationship between this and symptoms of respiratory disease is less well described. Charles Blackley (1880) was probably the first to suggest an association between asthma and fungal spore exposure. A number of investigators have explored the association between sensitisation to fungi, exposure, and asthma severity and exacerbations; fewer have investigated the link between fungal sensitisation and the onset of respiratory allergic disease. This section will briefly explore what is known about these associations.

9.3.2.1 Reported Associations between Asthma and Fungi

Allergic sensitisation to fungi is an important risk factor for allergic asthma, and fungal exposure has been linked to asthma exacerbations and hospital presentations (Tham *et al.*, 2014) as well as a described association with asthma mortality. Only a few of these most ecologically diverse organisms have been studied and linked to allergic respiratory disease.

Fungi have been suggested to be a trigger for development of asthma with genera such as *Alternaria*, *Aspergillus*, *Penicillium*, and *Cladosporium* associated with respiratory illness (Agarwal and Gupta, 2011; Denning *et al.*, 2006). This area of research has been limited by the ability to identify and quantify fungal species in a timely and cost-effective manner. A recent study (Reponen *et al.*, 2012) has used fungal-specific quantitative polymerase chain reaction to identify and quantify indoor fungal species, relating this data to development of asthma symptoms at age seven in a group of children whose indoor environment had been sampled in the first year of life and then again in their seventh year. Following multivariate analysis, there was a statistically higher risk of asthma at age seven in those children from homes with the highest indoor fungal levels, with the strongest association demonstrated for three species – *Aspergillus ochraceus*, *Aspergillus unguis*, and *Penicillium variabile* (Reponen *et al.*, 2012).

A small number of studies have described associations between fungal spore counts, allergic sensitisation, and asthma attacks (Lebowitz *et al.*, 1987; Targonski *et al.*, 1995). Targonski *et al.* (1995) described an association between fungal spore counts and asthma deaths by monitoring fungi and pollen concentrations using a Rotorod sampler and correlating these results with asthma mortality data in Chicago. Asthma deaths were observed among people 5–34 years of age without a striking seasonal pattern. They found that fungal spore levels were significantly higher on days where asthma-related death occurred than on days with no deaths. This association remained valid on multivariate logistic regression controlling for pollen levels, and the odds of asthma death occurring was found to be 1.2 times higher for every increase of 1,000 fungal spores m^{-3} daily counts (Targonski *et al.*, 1995).

Asthma exacerbations commonly lead to primary care or ED presentations with associated implications for overall burden of asthma on the community and the individual. Over half of such presentations occur in children and adolescents and can be triggered by a variety of factors including respiratory viral infection, thunderstorms, pollutants, and aeroallergen exposure including pollen and fungi, the last of which is least studied. A recent review (Tham *et al.*, 2014) has investigated the existing literature for reports on this association and identified this factor as one deserving greater study. While a number of studies report an association, there are methodological flaws in most of them (Peden and Reed, 2010). Pongracic *et al.* (2010) reported on research on associations between both indoor and outdoor fungal levels and increases in asthma symptoms and exacerbations occasioning hospital visits, finding indoor exposure to *Penicillium* linked to asthma exacerbations.

A large study across ten Canadian cities (Dales *et al.*, 2004) found consistent associations between asthma hospitalisations and fungal spore levels but differences in the type of fungi responsible, with Basidiomycetes having the strongest effect followed by Deuteromycetes and Ascomycetes.

9.3.2.2 Indoor Dampness/Mould

There is a growing body of evidence for an association between indoor dampness and fungi and a variety of allergic and non-allergic effects – the former including asthma (diagnosis, exacerbation, current), AR, and eczema, and the latter including cough, bronchitis, and increased respiratory infections. The *Damp Indoor Spaces and Health* report (Institute of Medicine, 2004), contains a review of epidemiological evidence on the health effects of damp housing, finding no evidence of a causal relationship but evidence of an association for cough, wheeze, asthma exacerbations, upper respiratory tract symptoms, and hypersensitivity pneumonitis. Mendell *et al.* (2011) have recently published an updated review of the literature (to 2009), finding consistent positive associations between dampness, mould, and a variety of respiratory symptoms. Evident dampness and mould and its association with respiratory illness may represent direct relationships between these symptoms and levels of fungi within the environment, but so far, accurate measurements of fungal levels have been difficult and complex. Strongest evidence exists for a causal relationship between dampness and mould and asthma exacerbations. Kercsmar *et al.* (2006) conducted an interventional study with children having frequent exacerbations and showed dramatic reductions in asthma presentations once remediation work reduced exposure to dampness and mould.

A number of studies have examined the relationship between asthma development and indoor dampness or mould. This literature has been summarised by Mendell *et al.* (2011). Pekkanen *et al.* (2007) showed that the development of asthma in young children was related, in a dose-dependent manner, to dampness and mould in the main areas of the home.

Both Institute of Medicine (2004) and the systematic review by Mendell *et al.* (2011) report that the available literature consistently supports an association between doctor-diagnosed AR and dampness and mould, with odds ratios ranging from 0.7 to 3.5. A dose–response relationship between the quantity of visible mould and increases in AR was found in a prospective study (Biagini *et al.*, 2006).

Rising levels of climate-altering pollutants such as CO_2 are known to favour production and release of pollen (Chapter 2) and, therefore, may contribute to increasing rates of respiratory allergy (Beggs and Bambrick, 2005). Until recently, little has been known about possible effects of increasing CO_2 on fungal allergen production. Wolf *et al.* (2010) studied this and demonstrated the increased plant biomass and carbon-to-nitrogen ratio resulting from higher CO_2 levels cause an increase in *Alternaria* spore production and allergen content. Thus, just as for certain pollen species, allergenic fungi are likely to be amplified as CO_2 rises and are likely to contribute to increasing allergic symptoms.

9.4 Extreme Weather Events: Thunderstorms, Dust Storms, and Hurricane Katrina

Extreme weather events are projected to become more common as changes to our climate take hold (Chapter 1). Therefore, understanding the possible effects such occurrences may have on those with allergic respiratory conditions will be important in managing possible periods of increased attendances in EDs because of triggering acute asthma and other respiratory symptoms.

Hurricane Katrina and its aftermath, various reported thunderstorm events, and dust storms, are just a few examples of extreme weather events that have significant impacts on respiratory allergic disease causation and exacerbation.

9.4.1 Hurricane Katrina

Studies after Hurricane Katrina in August 2005 have provided further insights into the effects of fungi on health. During that time, several levees surrounding New Orleans were breached, resulting in >75% of the city being under water. Homes were inundated with floodwaters for weeks, resulting in profuse mould and bacterial growth. There were reports of a persistent cough in many residents – 'Katrina cough' – thought to be due to higher mould and particulate counts (Manuel, 2006). Hospitals reported increases in numbers of patients being seen for respiratory and cold symptoms. Rates of children with asthma in the New Orleans area were reported to increase from 14% in 2003 to 18% in 2006 (Epstein and Mills, 2005).

9.4.2 Thunderstorm Asthma

In Birmingham, UK, in 1983 (Packe and Ayres, 1985), the number of asthma ED presentations on 2 days in July increased fivefold after a heavy downfall of rain. Subsequent investigation found that this event was associated with a marked increase in fungal spores, particularly *Didymella* and *Sporobolomyces*. Since that time, thunderstorms have been linked to asthma epidemics in various parts of the world, including London, Ontario, Atlanta, Melbourne, and other parts of Australia (Dabrera *et al.*, 2013).

In 1987, major hospitals in Melbourne reported a five- to ten-fold increase in asthma admissions during a 24-hour period that was characterised by increased rainfall and humidity, sudden drop in temperature, and no increase in atmospheric pollutants. Aeroallergens were not commented upon in this report (Bellomo *et al.*, 1992).

Dales *et al.* (2003) compared rates of asthma ED admissions to a children's hospital in Ontario on days with and without thunderstorms. These observations were accompanied by measurements of exposure to pollen and mould spore aeroallergens, as well as measures of various air pollutants. They found that daily asthma admissions increased 15% on thunderstorm days and that spore levels doubled on these days. Grass pollen was not observed to increase over this period. This data was consistent with their previous observation (Dales *et al.*, 2000) that asthma admissions increased on days with high spore counts irrespective of thunderstorm activity.

While an increase in grass pollen allergen levels has been hypothesised as an explanation for thunderstorm asthma, this has not been uniformly supported by the data. Wetting pollen grains can cause their rupture with release of small, respirable starch granules containing allergen, and these pollen fragments are then deposited close to ground level where sudden high-level exposure for sensitised individuals leads to symptoms (Dabrera *et al.*, 2013). Schäppi *et al.* (1999) demonstrated an increase in grass group 5 allergens in the atmosphere on days with light rainfall compared to dry days. Lewis *et al.* (2000) demonstrated an effect of pollen levels on asthma admissions on days with light rainfall, while they found no effect on dry days. Others have observed an association between asthma admissions, thunderstorm activity, and pollen counts only when thunderstorms follow several days of high pollen counts (Taylor and Jonsson, 2004). Grundstein *et al.* (2008) described an association between asthma ED visits and thunderstorm activity in Atlanta, relating the increases in such visits to meteorological factors such as rainfall and wind speed.

A number of other explanations have been given for the phenomenon of 'thunderstorm asthma'. Thunderstorm activity is characterised by strong downdrafts and cold dry outflows. There is some evidence that certain fungi rapidly increase in numbers associated with thunderstorm activity. Turbulent winds may increase spore release as well as recirculate deposited spores, making them available for inhalation (Dales *et al.*, 2003). Finally, others have demonstrated rises in sulphur dioxide concentrations as a causal agent (Celenza *et al.*, 1996).

9.4.3 Dust Storms and Long-Distance Transport

Over the past 30 years, there has been a significant increase in the transport of dust across the Atlantic and Pacific Oceans. There is evidence that this phenomenon has doubled since the 1970s. Drought conditions in north Africa have led to greater dust generation and transport. Dust generated in Asia can be associated with biological aerosols and pollutant particulate matter before crossing the Pacific Ocean (Shinn *et al.*, 2003).

Asthma has increased dramatically in the Caribbean (Monteil *et al.*, 2000). The Caribbean Asthma Association reported a seventeen-fold increase in asthma on the island of Barbados since 1973, which corresponds to a prevalence rate of approximately one in four inhabitants. In the same time period, there has been a dramatic increase in Saharan soil dust transport to Miami and Barbados (Prospero *et al.*, 2008).

From time to time, Taiwan is affected by yellow dust storms blown in from China. Wu *et al.* (2004) used a Burkard spore trap to monitor atmospheric spores, comparing dust storm day levels with those on other days during 2000–2001. The composition of dominant spores such as Basidiospore, *Penicillium*, *Aspergillus*, *Nigrospora*, *Arthrinium*, *Curvularia*, rusts, *Stemphylium*, *Cercospora*, *Pithomyces*, and unidentified fungi were significantly higher than those of background days (Wu *et al.*, 2004). The increase in Basidiospore, *Penicillium/Aspergillus*, *Nigrospora*, and unidentified fungi was significantly associated with the increase of ambient particulate levels with regression coefficients ranging from 0.887 to 31.98 (Wu *et al.*, 2004).

9.5 Pollutants, Aeroallergens, and Allergic Disease

Urbanisation and global industrialisation have resulted in an increase in airborne particulate matter, in particular DEPs, as well as gaseous compounds; various weather conditions influence their airborne levels and distribution (Smith *et al.*, 2014; Chapter 8). Ambient air pollutants have been shown to impact negatively on several health outcomes, and in particular, they have a complex relationship with allergies and asthma. A few of the many publications addressing these issues are summarised in the following sections.

9.5.1 Link with Increasing Sensitisation to Allergen

A Japanese group were the first to report on the phenomenon of greater pollen-induced AR in urban compared to rural dwellers where intuitively one would expect the reverse situation. They found greater Japanese cedar pollinosis in residents living close to urban highways when compared to residents living in close proximity to the cedar forests in rural locations. Further investigations revealed that air pollutants may interact with aeroallergens, taking on the role of adjuvants and leading to greater sensitisation to the aeroallergens as well as exacerbation of symptoms in already sensitised individuals (Ishizaki *et al.*, 1987).

9.5.2 Link with Increase in Inflammatory Markers

There is evidence suggesting that ozone increases airway inflammation and permeability of the bronchial epithelial surface. In asthmatic subjects, increased cells, neutrophils in particular, and inflammatory mediators are found in lavage fluid after ozone exposure. These findings provide a mechanism explaining the observation of increased bronchial hyperreactivity and decrease in lung function in subjects with asthma after ozone exposure (Riedl and Diaz-Sanchez, 2005). DEPs are also known to cause significant airway inflammation and have been shown to produce an adjuvant effect with pollen exposure (Riedl and Diaz-Sanchez, 2005).

9.5.3 Link with Lung Development in Children

An 8-year longitudinal study of 3,600 children living close to highways in California demonstrated a lack of lung function growth when compared to children living at a greater distance from these roadways (Gauderman *et al.*, 2007).

9.5.4 Link with New-Onset Asthma

In a 5-year childhood cohort study using participation in sports as a marker for time spent outdoors, McConnell *et al.* (2002) demonstrated that the relative risk of asthma development in children playing three or more sports was 3.3 (95% confidence interval, 1.9–5.8) compared with children playing no sports, in communities with exposure to high ozone concentration. An effect of sport on asthma development was not seen in communities with low ozone concentration.

9.5.5 ED Presentations

A number of studies have provided evidence for a link between exposure to traffic-related air pollutants such as nitrogen dioxide (NO_2) and ozone and need for hospital admission with an asthma exacerbation (D'Amato and Cecchi, 2008). This link appears most pronounced in childhood and young adulthood. Studies linking ozone exposure to increased ED presentations for asthma exacerbations have been published for Atlanta (White *et al.*, 1994) and Mexico City (Romieu *et al.*, 1995). In a study in Birmingham, Edwards *et al.* (1994) demonstrated that, compared to children admitted for non-respiratory diagnoses, children admitted for an asthma exacerbation were more likely to live near major roads in areas of high traffic levels. In Germany, Nicolai *et al.* (2003) provided evidence for a link between increased traffic pollutant exposure and incidence of asthma, cough, and wheeze in a large survey of school children.

Villeneuve *et al.* (2007) examined the effects of outdoor air pollution on ED visits for asthma in children and young adults in Edmonton, Canada. They found strong associations for NO_2 and carbon monoxide (CO) levels at certain times of the year and associations also with ozone and particulate matter. These and many other studies demonstrate complex and significant relationships between various air pollutants and many aspects of the allergic response and respiratory disease.

9.6 Influences on Other Allergic Conditions

This chapter has dealt with allergic respiratory disorders, asthma in particular, but it is highly likely that other allergic disorders are being and will continue to be impacted by climate change and its downstream effects.

9.6.1 Atopic Dermatitis

AD is a common chronic skin condition that is an early manifestation of the atopic march. This is a costly condition requiring regular emollient use, topical corticosteroid creams, and some oral treatments. US figures suggest that up to 20% children and 2% to 10% adults are affected (Mancini *et al.*, 2008).

Along with the other atopic conditions, the prevalence of AD has increased two- to three-fold in industrialised countries, and this can be linked to aspects of the Western lifestyle. A systematic review of the impact of AD in the United States suggested that approximately US$ 3.8 billion is spent annually on physician visits and medications (Mancini *et al.*, 2008).

Aeroallergen exposure is an important factor driving or exacerbating AD for a subset of patients. HDM and cockroach proteins are probably the most important allergens, with skin contact leading to onset or flares in AD skin. These proteins may act in a number of ways to disturb the skin barrier and cause acute lesions. First, some HDM proteins have innate proteolytic activity and may disrupt epithelial tight junctions, activate a number of cell populations, and stimulate the production of proinflammatory cytokines resulting in increased local inflammation and disruption to the skin barrier. This may be complicated by easier access through the epidermis of a number of irritant and infective agents driving further initiation of allergic and inflammatory responses (Nakamura *et al.*, 2006).

Airborne proteins may directly activate protease-activated receptor 2 (PAR2) receptors in the skin, promoting itching and scratching that further leads to disruption of the epithelial barrier. PAR2 receptors are essential for itch transmission in the skin (Jeong *et al.*, 2008).

As with respiratory allergy, aeroallergens may stimulate IgE-mediated responses resulting in histamine release, increased itching, and allergic inflammation that causes further flaring and barrier disruption. Thus, changes in climatic factors that impact on distribution and burden of HDM, cockroach, pollen, and mould aeroallergens (Chapters 2–8) are likely to exert an effect on the severity and possibly the frequency of AD (Hostetler *et al.*, 2010).

9.6.2 Insect Sting Allergy

Hymenoptera venom allergy caused by an IgE-mediated allergic reaction is responsible for significant morbidity and adversely impacts quality of life in those at risk of a systemic reaction. Stinging insect allergy is a public health concern because of the high frequency of insect stings and prevalence of life-threatening systemic allergic reactions and death. Stinging insect allergy is often seen in certain occupational settings such as in gardeners, groundsmen, linesmen, and many other outdoor occupations.

The prevalence of large local reactions to stings ranges from 2.4% to 26.4% in the general population and up to 38% in bee keepers. The prevalence of allergic systemic reactions is between 0.3% and 7.5% in Europe, while in the United States, it is 0.5–3.3%. Allergic systemic reactions are less common in children, ranging from 0.15–0.8% (World Allergy Organization, 2011).

A higher risk of stings is dependent on geographic location, climate, temperature, insect behaviour, occupation, leisure and sporting activities, and proximity of hives and nests to dwellings and the workplace. Projected changes in vegetation, climate, and temperature will almost certainly impact on the distribution of stinging insects (Chapter 3) and thus will influence the prevalence and pattern of sting reactions.

9.6.3 Food Allergy

Food allergies affect up to 8% of children and 3% to 4% of adults (Sicherer and Sampson, 2010). Up to 50% of all cases of anaphylaxis are caused by a food allergic reaction (Decker *et al.*, 2008), with peanut being the commonest cause of fatal food-induced anaphylaxis. There is good evidence that the prevalence of food allergy is increasing. In the last decade, reported food allergy prevalence increased by 18% in children under 18 years; peanut allergy has increased from 0.4% in 1997 to 1.4% in 2007.

IgE-mediated food allergies may result from sensitisation to food allergens (class I food allergens) via the gastrointestinal tract or may occur as a result of primary sensitisation to homologous pollen allergens via the respiratory tract causing reactivity to cross-reactive food allergens (class II allergens). Global differences in food sensitisation patterns have been particularly observed for these plant food allergens whereby differences in allergenic plant distribution, agriculture, and dietary patterns determine the predominant pattern of pollen and food allergy. For instance, in Europe, prevalence of plant food allergy is significantly influenced by sensitisation to particular proteins in birch pollen such as Bet v 1 and Bet v 2, while in the Mediterranean region, there is a higher sensitisation rate to profilins and non-specific lipid transfer proteins (Burney *et al.*, 2014). A recent study has shown differences in the pattern of allergen reactivity causing peanut sensitisation across different geographic regions, and these differences are largely determined by aeroallergen exposure (Vereda *et al.*, 2011). Thus, it is likely that changes in climate that result in altered distribution of various allergenic plants (Chapter 3) may in time bring about a change in the pattern of food allergy, especially that caused by plant food allergens.

9.7 Unmet Needs, Future Research Directions

There are numerous limitations to our understanding of the complex interactions between allergic diseases and environmental factors such as pollen and spore levels, climatic factors, and air pollutants. There are conflicting data from epidemiological studies, and some regions of the world have been well studied while nothing exists for other parts. There is limited longitudinal data in existence that may assist in understanding relationships, and given the nature of the data, they cannot be extrapolated from one region to another. Traditional methods of elaborating pollen and spore counts do not reflect totally the airborne allergen load that an individual may be exposed to at a given time. Most importantly, we have very little knowledge of threshold levels for sensitisation and for symptom elicitation whether we consider pollen, mould, or ambient pollutants.

Many of these problems are difficult to address, but from what has already been documented and from the changes projected to occur as a result of changes in our climate (Chapters 1–8), it is imperative that these gaps in our knowledge be addressed and that we continue to establish improved methods of monitoring the various parameters that will certainly be impacted by global climate change.

Establishment of long-term aerobiological networks and pollution monitoring for every country must be a goal as we monitor effects of climate change on respiratory health. This will necessitate the development of more cost-effective, portable, and accurate instrumentation with which to gather this data. Allergic patient population data will be needed to correlate aerobiological effects with health outcomes. In addition, projected climatic changes are likely to impact on other forms of allergic disease such as food allergy, AD, and stinging insect allergy.

The last 20 years has seen a significant rise in all allergic disorders leading some to describe the present situation as an 'allergy epidemic'. The currently available medical workforce is not meeting the needs of allergic patients in any region, so with the likely changes and increases in some allergic disorders, it is imperative that measures to train and upskill medical workforces in each region be undertaken as a healthcare priority.

References

Agarwal, R., Gupta, D. (2011). Severe asthma and fungi: current evidence. *Medical Mycology*, 49(Suppl 1), S150–S157.

Akdis, C. A., Agache, I., eds. (2013). *Global Atlas of Asthma*. Zurich: European Academy of Allergy and Clinical Immunology.

Ariano, R., Canonica, G. W., Passalacqua, G. (2010). Possible role of climate changes in variations in pollen seasons and allergic sensitizations during 27 years. *Annals of Allergy, Asthma & Immunology*, 104(3), 215–222.

Asher, M. I., Montefort, S., Björkstén, B., *et al.* (2006). Worldwide time trends in the prevalence of symptoms of asthma, allergic rhinoconjunctivitis, and eczema in childhood: ISAAC Phases One and Three repeat multicountry cross-sectional surveys. *The Lancet*, 368(9537), 733–743.

Babin, S. M., Burkom, H. S., Holtry, R. S., *et al.* (2007). Pediatric patient asthma-related emergency department visits and admissions in Washington, DC, from 2001–2004, and associations with air quality, socio-economic status and age group. *Environmental Health*, 6, 9.

Bahadori, K., Doyle-Waters, M. M., Marra, C., *et al.* (2009). Economic burden of asthma: a systematic review. *BMC Pulmonary Medicine*, 9, 24.

Beggs, P. J. (2004). Impacts of climate change on aeroallergens: past and future. *Clinical and Experimental Allergy*, 34(10), 1507–1513.

Beggs, P. J., Bambrick, H. J. (2005). Is the global rise of asthma an early impact of anthropogenic climate change? *Environmental Health Perspectives*, 113(8), 915–919.

Bellomo, R., Gigliotti, P., Treloar, A., *et al.* (1992). Two consecutive thunderstorm associated epidemics of asthma in the city of Melbourne. The possible role of rye grass pollen. *The Medical Journal of Australia*, 156(12), 834–837.

Biagini, J. M., LeMasters, G. K., Ryan, P. H., *et al.* (2006). Environmental risk factors of rhinitis in early infancy. *Pediatric Allergy and Immunology*, 17(4), 278–284.

Bielory, L., Friedlaender, M. H. (2008). Allergic conjunctivitis. *Immunology and Allergy Clinics of North America*, 28(1), 43–58.

Blackley, C. H. (1880). *Hay Fever: Its Causes, Treatment, and Effective Prevention. Experimental Researches*, 2nd edn. London: Baillière, Tindall, & Cox.

Borish, L. (2003). Allergic rhinitis: systemic inflammation and implications for management. *The Journal of Allergy and Clinical Immunology*, 112(6), 1021–1031.

Breton, M.-C., Garneau, M., Fortier, I., Guay, F., Louis, J. (2006). Relationship between climate, pollen concentrations of *Ambrosia* and medical consultations for allergic rhinitis in Montreal, 1994–2002. *Science of the Total Environment*, 370(1), 39–50.

Burney, P., Chinn, S., Jarvis, D., *et al.* (1996).Variations in the prevalence of respiratory symptoms, self-reported asthma attacks, and use of asthma medication in the European Community Respiratory Health Survey (ECRHS). *European Respiratory Journal*, 9(4), 687–695.

Burney, P. G. J., Potts, J., Kummeling, I., *et al.* (2014). The prevalence and distribution of food sensitization in European adults. *Allergy*, 69(3), 365–371.

Burr, M. L., Emberlin, J. C., Treu, R., *et al.* (2003). Pollen counts in relation to the prevalence of allergic rhinoconjunctivitis, asthma and atopic eczema in the International Study of Asthma and Allergies in Childhood (ISAAC). *Clinical and Experimental Allergy*, 33(12), 1675–1680.

Carlsen, K. H., Ørstavik, I., Leegaard, J., Høeg, H. (1984). Respiratory virus infections and aeroallergens in acute bronchial asthma. *Archives of Disease in Childhood*, 59(4), 310–315.

Cecchi, L., D'Amato, G., Ayres, J. G., *et al.* (2010). Projections of the effects of climate change on allergic asthma: the contribution of aerobiology. *Allergy*, 65(9), 1073–1081.

Celenza, A., Fothergill, J., Kupek, E., Shaw, R. J. (1996). Thunderstorm associated asthma: a detailed analysis of environmental factors. *BMJ-British Medical Journal*, 312(7031), 604–607.

Charpin, D., Hughes, B., Mallea, M., *et al.* (1993). Seasonal allergic symptoms and their relation to pollen exposure in south-east France. *Clinical and Experimental Allergy*, 23(5), 435–439.

Chipps, B. E., Zeiger, R. S., Borish, L., *et al.* (2012). Key findings and clinical implications from The Epidemiology and Natural History of Asthma: Outcomes and Treatment Regimens (TENOR) study. *The Journal of Allergy and Clinical Immunology*, 130(2), 332–342.e10.

Craig, T. J. (2010). Aeroallergen sensitization in asthma: prevalence and correlation with severity. *Allergy and Asthma Proceedings*, 31(2), 96–102.

Crystal-Peters, J., Crown, W. H., Goetzel, R. Z., Schutt, D. C. (2000). The cost of productivity losses associated with allergic rhinitis. *The American Journal of Managed Care*, 6(3), 373–378.

Dabrera, G., Murray, V., Emberlin, J., *et al.* (2013). Thunderstorm asthma: an overview of the evidence base and implications for public health advice. *QJMed-An International Journal of Medicine*, 106(3), 207–217.

Dales, R. E., Cakmak, S., Burnett, R. T., Judek, S., Coates, F., Brook, J. R. (2000). Influence of ambient fungal spores on emergency visits for asthma to a regional children's hospital. *American Journal of Respiratory and Critical Care Medicine*, 162(6), 2087–2090.

Dales, R. E., Cakmak, S., Judek, S., *et al.* (2003). The role of fungal spores in thunderstorm asthma. *Chest*, 123(3), 745–750.

Dales, R. E., Cakmak, S., Judek, S., *et al.* (2004). Influence of outdoor aeroallergens on hospitalization for asthma in Canada. *The Journal of Allergy and Clinical Immunology*, 113(2), 303–306.

D'Amato, G., Cecchi, L. (2008). Effects of climate change on environmental factors in respiratory allergic diseases. *Clinical and Experimental Allergy*, 38(8), 1264–1274.

Darrow, L. A., Hess, J., Rogers, C. A., Tolbert, P. E., Klein, M., Sarnat, S. E. (2012). Ambient pollen concentrations and emergency department visits for asthma and wheeze. *The Journal of Allergy and Clinical Immunology*, 130(3), 630–638.e4.

Decker, W. W., Campbell, R. L., Manivannan, V., *et al.* (2008). The etiology and incidence of anaphylaxis in Rochester, Minnesota: a report from the Rochester Epidemiology Project. *The Journal of Allergy and Clinical Immunology*, 122(6), 1161–1165.

Denning, D. W., O'Driscoll, B. R., Hogaboam, C. M., Bowyer, P., Niven, R. M. (2006). The link between fungi and severe asthma: a summary of the evidence. *European Respiratory Journal*, 27(3), 615–626.

Edwards, J., Walters, S., Griffiths, R. K. (1994). Hospital admissions for asthma in preschool children: relationship to major roads in Birmingham, United Kingdom. *Archives of Environmental Health*, 49(4), 223–227.

Epstein, P. R., Mills, E., eds. (2005). *Climate Change Futures: Health, Ecological and Economic Dimensions*. Boston: The Center for Health and the Global Environment, Harvard Medical School.

Erbas, B., Akram, M., Dharmage, S. C., *et al.* (2012). The role of seasonal grass pollen on childhood asthma emergency department presentations. *Clinical & Experimental Allergy*, 42(5), 799–805.

Fitter, A. H., Fitter, R. S. R. (2002). Rapid changes in flowering time in British plants. *Science*, 296(5573), 1689–1691.

Galán, I., Prieto, A., Rubio, M., *et al.* (2010). Association between airborne pollen and epidemic asthma in Madrid, Spain: a case–control study. *Thorax*, 65(5), 398–402.

Gauderman, W. J., Vora, H., McConnell, R., *et al.* (2007). Effect of exposure to traffic on lung development from 10 to 18 years of age: a cohort study. *The Lancet*, 369(9561), 571–577.

Grundstein, A., Sarnat, S. E., Klein, M., *et al.* (2008). Thunderstorm associated asthma in Atlanta, Georgia. *Thorax*, 63(7), 659–660.

Guerra, S., Sherrill, D. L., Martinez, F. D., Barbee, R. A. (2002). Rhinitis as an independent risk factor for adult-onset asthma. *The Journal of Allergy and Clinical Immunology*, 109(3), 419–425.

Harley, K. G., Macher, J. M., Lipsett, M., *et al.* (2009). Fungi and pollen exposure in the first months of life and risk of early childhood wheezing. *Thorax*, 64(4), 353–358.

Héguy, L., Garneau, M., Goldberg, M. S., Raphoz, M., Guay, F., Valois, M.-F. (2008). Associations between grass and weed pollen and emergency department visits for asthma among children in Montreal. *Environmental Research*, 106(2), 203–211.

Hostetler, S. G., Kaffenberger, B., Hostetler, T., Zirwas, M. J. (2010). The role of airborne proteins in atopic dermatitis. *The Journal of Clinical and Aesthetic Dermatology*, 3(1), 22–31.

Institute of Medicine (2004). *Damp Indoor Spaces and Health.* Washington, DC: The National Academies Press.

Ishizaki, T., Koizumi, K., Ikemori, R., Ishiyama, Y., Kushibiki, E. (1987). Studies of prevalence of Japanese cedar pollinosis among the residents in a densely cultivated area. *Annals of Allergy*, 58(4), 265–270.

Jarvis, D., Newson, R., Lotvall, J., *et al.* (2012). Asthma in adults and its association with chronic rhinosinusitis: the GA^{2}LEN survey in Europe. *Allergy*, 67(1), 91–98.

Jeong, S. K., Kim, H. J., Youm, J.-K., *et al.* (2008). Mite and cockroach allergens activate protease-activated receptor 2 and delay epidermal permeability barrier recovery. *Journal of Investigative Dermatology*, 128(8), 1930–1939.

Katelaris, C. H. (2003). Ocular allergy: implications for the clinical immunologist. *Annals of Allergy, Asthma & Immunology*, 90(6 Suppl), 23–27.

Kercsmar, C. M., Dearborn, D. G., Schluchter, M., *et al.* (2006). Reduction in asthma morbidity in children as a result of home remediation aimed at moisture sources. *Environmental Health Perspectives*, 114(10), 1574–1580.

Kim, H., Bernstein, J. A. (2009). Air pollution and allergic disease. *Current Allergy and Asthma Reports*, 9(2), 128–133.

Kulig, M., Klettke, U., Wahn, V., *et al.* (2000). Development of seasonal allergic rhinitis during the first 7 years of life. *The Journal of Allergy and Clinical Immunology*, 106(5), 832–839.

Lebowitz, M. D., Collins, L., Holberg, C. J. (1987). Time series analyses of respiratory responses to indoor and outdoor environmental phenomena. *Environmental Research*, 43(2), 332–341.

Lewis, S. A., Corden, J. M., Forster, G. E., Newlands, M. (2000). Combined effects of aerobiological pollutants, chemical pollutants and meteorological conditions on asthma admissions and A & E attendances in Derbyshire UK, 1993–96. *Clinical and Experimental Allergy*, 30(12), 1724–1732.

Lierl, M. B., Hornung, R. W. (2003). Relationship of outdoor air quality to pediatric asthma exacerbations. *Annals of Allergy, Asthma & Immunology*, 90(1), 28–33.

Mancini, A. J., Kaulback, K., Chamlin, S. L. (2008). The socioeconomic impact of atopic dermatitis in the United States: a systematic review. *Pediatric Dermatology*, 25(1), 1–6.

Manuel, J. (2006). In Katrina's wake. *Environmental Health Perspectives*, 114(1), A32–A39. (See correction in: Errata (2006). *Environmental Health Perspectives*, 114(2), A90.)

Masoli, M., Fabian, D., Holt, S., Beasley, R., for the Global Initiative for Asthma (GINA) Program (2004). The global burden of asthma: executive summary of the GINA Dissemination Committee Report. *Allergy*, 59(5), 469–478.

McConnell, R., Berhane, K., Gilliland, F., *et al.* (2002). Asthma in exercising children exposed to ozone: a cohort study. *The Lancet*, 359(9304), 386–391. (See correction in: Department of Error (2002). *The Lancet*, 359(9309), 896.)

Mendell, M. J., Mirer, A. G., Cheung, K., Tong, M., Douwes, J. (2011). Respiratory and allergic health effects of dampness, mold, and dampness-related agents: a review of the epidemiologic evidence. *Environmental Health Perspectives*, 119(6), 748–756.

Menzel, A. (2000). Trends in phenological phases in Europe between 1951 and 1996. *International Journal of Biometeorology*, 44(2), 76–81.

Monteil, M. A., Juman, S., Hassanally, R., *et al.* (2000). Descriptive epidemiology of asthma in Trinidad, West Indies. *Journal of Asthma*, 37(8), 677–684.

Nakamura, T., Hirasawa, Y., Takai, T., *et al.* (2006). Reduction of skin barrier function by proteolytic activity of a recombinant house dust mite major allergen Der f 1. *Journal of Investigative Dermatology*, 126(12), 2719–2723.

Nathan, R. A. (2007). The burden of allergic rhinitis. *Allergy and Asthma Proceedings*, 28(1), 3–9.

Nicolai, T., Carr, D., Weiland, S. K., *et al.* (2003). Urban traffic and pollutant exposure related to respiratory outcomes and atopy in a large sample of children. *European Respiratory Journal*, 21(6), 956–963.

Ogershok, P. R., Warner, D. J., Hogan, M. B., Wilson, N. W. (2007). Prevalence of pollen sensitization in younger children who have asthma. *Allergy and Asthma Proceedings*, 28(6), 654–658.

Ozdoganoglu, T., Songu, M. (2012). The burden of allergic rhinitis and asthma. *Therapeutic Advances in Respiratory Disease*, 6(1), 11–23.

Packe, G. E., Ayres, J. G. (1985). Asthma outbreak during a thunderstorm. *The Lancet*, 326(8448), 199–204.

Peden, D., Reed, C. E. (2010). Environmental and occupational allergies. *The Journal of Allergy and Clinical Immunology*, 125(2 Suppl 2), S150–S160.

Pekkanen, J., Hyvärinen, A., Haverinen-Shaughnessy, U., *et al.* (2007). Moisture damage and childhood asthma: a population-based incident case-control study. *European Respiratory Journal*, 29(3), 509–515.

Pongracic, J. A., O'Connor, G. T., Muilenberg, M. L., *et al.* (2010). Differential effects of outdoor versus indoor fungal spores on asthma morbidity in inner-city children. *The Journal of Allergy and Clinical Immunology*, 125(3), 593–599.

Prospero, J. M., Blades, E., Naidu, R., *et al.* (2008). Relationship between African dust carried in the Atlantic trade winds and surges in pediatric asthma attendances in the Caribbean. *International Journal of Biometeorology*, 52(8), 823–832.

Rastogi, D., Reddy, M., Neugebauer, R. (2006). Comparison of patterns of allergen sensitization among inner-city Hispanic and African American children with asthma. *Annals of Allergy, Asthma & Immunology*, 97(5), 636–642.

Reid, C. E., Gamble, J. L. (2009). Aeroallergens, allergic disease, and climate change: impacts and adaptation. *EcoHealth*, 6(3), 458–470.

Reponen, T., Lockey, J., Bernstein, D. I., *et al.* (2012). Infant origins of childhood asthma associated with specific molds. *The Journal of Allergy and Clinical Immunology*, 130(3), 639–644.

Riedl, M., Diaz-Sanchez, D. (2005). Biology of diesel exhaust effects on respiratory function. *The Journal of Allergy and Clinical Immunology*, 115(2), 221–228.

Romieu, I., Meneses, F., Sienra-Monge, J. J. L., *et al.* (1995). Effects of urban air pollutants on emergency visits for childhood asthma in Mexico City. *American Journal of Epidemiology*, 141(6), 546–553.

Rossi, O. V. J., Kinnula, V. L., Tienari, J., Huhti, E. (1993). Association of severe asthma attacks with weather, pollen, and air pollutants. *Thorax*, 48(3), 244–248.

Schäppi, G. F., Taylor, P. E., Pain, M. C. F., *et al.* (1999). Concentrations of major grass group 5 allergens in pollen grains and atmospheric particles: implications for hay fever and allergic asthma sufferers sensitized to grass pollen allergens. *Clinical and Experimental Allergy*, 29(5), 633–641.

Schmier, J. K., Ebi, K. L. (2009). The impact of climate change and aeroallergens on children's health. *Allergy and Asthma Proceedings*, 30(3), 229–237.

Sembajwe, G., Cifuentes, M., Tak, S. W., *et al.* (2010). National income, self-reported wheezing and asthma diagnosis from the World Health Survey. *European Respiratory Journal*, 35(2), 279–286.

Settele, J., Scholes, R., Betts, R. A., *et al.* (2014). Terrestrial and inland water systems. In: Field, C. B., Barros, V. R., Dokken, D. J., *et al.*, eds. *Climate Change 2014: Impacts, Adaptation, and Vulnerability. Part A: Global and Sectoral Aspects. Contribution of Working Group II to the Fifth Assessment Report of the Intergovernmental Panel on Climate Change*. Cambridge, UK and New York, NY: Cambridge University Press, pp. 271–359.

Shea, K. M., Truckner, R. T., Weber, R. W., Peden, D. B. (2008). Climate change and allergic disease. *The Journal of Allergy and Clinical Immunology*, 122(3), 443–453.

Shinn, E. A., Griffin, D. W., Seba, D. B. (2003). Atmospheric transport of mold spores in clouds of desert dust. *Archives of Environmental Health*, 58(8), 498–504.

Sicherer, S. H., Sampson, H. A. (2010). Food allergy. *The Journal of Allergy and Clinical Immunology*, 125(2 Suppl 2), S116–S125.

Singer, B. D., Ziska, L. H., Frenz, D. A., Gebhard, D. E., Straka, J. G. (2005). Increasing Amb a 1 content in common ragweed (*Ambrosia artemisiifolia*) pollen as a function of rising atmospheric CO_2 concentration. *Functional Plant Biology*, 32(7), 667–670.

Smith, K. R., Woodward, A., Campbell-Lendrum, D., *et al.* (2014). Human health: impacts, adaptation, and co-benefits. In: Field, C. B., Barros, V. R., Dokken, D. J., *et al.*, eds. *Climate Change 2014: Impacts, Adaptation, and Vulnerability. Part A: Global and Sectoral Aspects. Contribution of Working Group II to the Fifth Assessment Report of the Intergovernmental Panel on Climate Change*. Cambridge, UK and New York, NY: Cambridge University Press, pp. 709–754.

Storkey, J., Stratonovitch, P., Chapman, D. S., Vidotto, F., Semenov, M. A. (2014). A process-based approach to predicting the effect of climate change on the distribution of an invasive allergenic plant in Europe. *PLoS One*, 9(2), e88156.

Szeinbach, S. L., Seoane-Vazquez, E. C., Beyer, A., Williams, P. B. (2007). The impact of allergic rhinitis on work productivity. *Primary Care Respiratory Journal*, 16(2), 98–105. (See correction in: Szeinbach, S. L. (2007). Correction. *Primary Care Respiratory Journal*, 16(4), 257.)

Takaro, T. K., Knowlton, K., Balmes, J. R. (2013). Climate change and respiratory health: current evidence and knowledge gaps. *Expert Review of Respiratory Medicine*, 7(4), 349–361.

Targonski, P. V., Persky, V. W., Ramekrishnan, V. (1995). Effect of environmental molds on risk of death from asthma during the pollen season. *The Journal of Allergy and Clinical Immunology*, 95(5 Pt 1), 955–961.

Taylor, P. E., Jonsson, H. (2004). Thunderstorm asthma. *Current Allergy and Asthma Reports*, 4(5), 409–413.

Tham, R., Dharmage, S. C., Taylor, P. E., *et al.* (2014). Outdoor fungi and child asthma health service attendances. *Pediatric Allergy and Immunology*, 25(5), 439–449.

Tobías, A., Galán, I., Banegas, J. R. (2004). Non-linear short-term effects of airborne pollen levels with allergenic capacity on asthma emergency room admissions in Madrid, Spain. *Clinical and Experimental Allergy*, 34(6), 871–878.

Traidl-Hoffmann, C., Kasche, A., Menzel, A., *et al.* (2003). Impact of pollen on human health: more than allergen carriers? *International Archives of Allergy and Immunology*, 131(1), 1–13.

Vereda, A., van Hage, M., Ahlstedt, S., *et al.* (2011). Peanut allergy: clinical and immunologic differences among patients from 3 different geographic regions. *The Journal of Allergy and Clinical Immunology*, 127(3), 603–607.

Villeneuve, P. J., Chen, L., Rowe, B. H., Coates, F. (2007). Outdoor air pollution and emergency department visits for asthma among children and adults: a case-crossover study in northern Alberta, Canada. *Environmental Health*, 6, 40.

Weber, R. W. (2012). Impact of climate change on aeroallergens. *Annals of Allergy, Asthma & Immunology*, 108(5), 294–299.

White, M. C., Etzel, R. A., Wilcox, W. D., Lloyd, C. (1994). Exacerbations of childhood asthma and ozone pollution in Atlanta. *Environmental Research*, 65(1), 56–68.

Wolf, J., O'Neill, N. R., Rogers, C. A., Muilenberg, M. L., Ziska, L. H. (2010). Elevated atmospheric carbon dioxide concentrations amplify *Alternaria alternata* sporulation and total antigen production. *Environmental Health Perspectives*, 118(9), 1223–1228.

World Allergy Organization (2011). *WAO White Book on Allergy*. Milwaukee: World Allergy Organization.

Wu, P.-C., Tsai, J.-C., Li, F.-C., Lung, S.-C., Su, H.-J. (2004). Increased levels of ambient fungal spores in Taiwan are associated with dust events from China. *Atmospheric Environment*, 38(29), 4879–4886.

Zhong, W., Levin, L., Reponen, T., *et al.* (2006). Analysis of short-term influences of ambient aeroallergens on pediatric asthma hospital visits. *Science of the Total Environment*, 370(2–3), 330–336.

Ziska, L. H., Caulfield, F. A. (2000). Rising CO_2 and pollen production of common ragweed (*Ambrosia artemisiifolia*), a known allergy-inducing species: implications for public health. *Australian Journal of Plant Physiology*, 27(10), 893–898.

Ziska, L., Knowlton, K., Rogers, C., *et al.* (2011). Recent warming by latitude associated with increased length of ragweed pollen season in central North America. *Proceedings of the National Academy of Sciences of the United States of America*, 108(10), 4248–4251.

10

Synthesis and Conclusion

PAUL J. BEGGS[1] AND LEWIS H. ZISKA[2]

[1]*Department of Environmental Sciences*
Faculty of Science and Engineering
Macquarie University
[2]*Crop Systems and Global Change*
Agricultural Research Service
US Department of Agriculture

10.1 Introduction

Every breath we take exposes us to a cocktail of gases and particles. Indeed, our interface with and therefore exposure to the atmospheric environment extends beyond the respiratory system, to also include our skin and our eyes. Aeroallergens are a component of this cocktail, and for the many in our community who suffer from allergic disease, this exposure can have serious health consequences.

Our climate is changing, and the impacts of this are well known and obvious in many instances – glaciers are shrinking, permafrost is thawing, species have shifted their geographic ranges and seasonal activities, etc. The impacts of climate change on allergens and allergic diseases are perhaps less obvious, or even hidden. They are, however, just as real. Indeed, the invisible nature of many environmental allergen sources – microscopic pollen grains, mould spores, and house dust mites – makes changes to them also invisible to most people. This makes people who suffer from allergic diseases highly vulnerable to any adverse impacts of climate change on allergens.

To the scientific climate change research community, impacts of climate change on allergens and allergic diseases are not a particularly new concept. A quarter of a century ago, Longstreth (1991) wrote, 'there are likely to be quantitative and/or qualitative changes in the airborne concentration of allergens, e.g., molds and pollens. This in turn could lead to changes in the prevalence or intensity of asthma and hay fever episodes in affected individuals'. Research since (observational, experimental, computational, and theoretical) has provided a wealth of evidence that such a statement is not only correct but that the nature of the impacts of climate change on allergens and allergic diseases extends way beyond, involving complexities that none of us could have even imagined 25 years ago.

The chapters of this book have elaborated on this complexity, each, in turn, focussing on a different aspect of the topic. The book as a whole, however, is not complete until a clear synthesis is provided, and it is this challenge that the following section addresses. The section following that highlights a range of knowledge gaps and research needs. Although not a focus of the book, this chapter also provides an overview of the basic responses to climate change impacts, specifically mitigation and adaptation, as well as adaptation responses specific to impacts on allergens and allergic diseases. This chapter, as well as the book, finishes with some words of encouragement and a call to action.

10.2 The Impacts of Climate Change on Allergens and Allergic Diseases

Climate change has had, is having, and will continue to have impacts on allergens. Climate influences the geographic distribution of all living organisms, including allergenic species, and with climate change comes not only the potential for movement of the boundaries of allergenic species but also increases/expansion or decreases/contraction in their overall range. The timing and duration of the presence of allergen in our environment is also climate-related, and we now know that climate change is changing when exposure to allergens may occur, in many cases advancing the start of spring including flowering and the production and release of pollen of certain plant species. Once allergens are airborne, their movement is determined by the direction and speed of the air that bares them, as well as the humidity of the air and the presence or absence of precipitation. All these factors are affected with climate change, with consequences for the dispersion, transport, and deposition of aeroallergens.

In addition to changes in when and where allergen exposure may occur, the quantity of allergen, the dose, is also changing in response to climate change. Observations at diverse global locations indicate that atmospheric pollen concentrations of a number of plant species are increasing, consistent with numerous controlled experiments showing that increasing temperatures and carbon dioxide (CO_2) concentrations enhance pollen production. In addition, these same factors – increasing air temperature and CO_2 concentration – may be resulting in increases in the allergen content, increasing pollen potency.

Unfortunately, our exposure to airborne allergens is often in combination with air pollutants, and this mix can augment the health consequences of allergen exposure. Air pollutants can also directly impact the allergen content of aeroallergens such as pollen, through changes in the plant during their formation, and interaction with such allergens in the atmosphere. While many of these impacts and interactions studied to date would have adverse implications for human health, for others the opposite is the case, so the overall health impact is difficult to quantify. In any case, all of this is changing, with climate change also having many impacts on air pollutants.

At present, our understanding of climate change impacts on allergens is best for pollen; however, the outdoor environment contains other important allergen sources including mould spores, stinging insects such as bees and wasps, and contact allergens such as poison ivy. Limited evidence suggests that they may also be impacted by climate change, but the extent of the impact is uncertain. The indoor environment introduces other allergen sources: cockroaches, mice, house dust mites, and certain fungi. Our homes, schools, offices, and other built environments do not disconnect us from the outdoor environment, including the outdoor climate. Extreme illustrations of this include the damage caused by tropical cyclones (hurricanes) and floods, and more subtle examples include the influence of the outdoor climate on the geographic distribution of these indoor organisms and seasonal variations in their activity. Among potential impacts of climate change, in regions for which increased precipitation is projected in the future, increased flooding will likely increase indoor fungal growth and associated allergen.

The prevalence of allergic diseases is rising dramatically in both developed and developing countries around the world and is already high, with hundreds of millions of people suffering from rhinitis and/or asthma (Pawankar *et al.*, 2011). The causes of this epidemic, or what Platts-Mills (2015) has recently referred to as 'the allergy epidemics', are a topic of continuing investigation and discussion. While the causes for the increase are many and complex, it is becoming increasingly clear that climate change is playing a role.

In this context, the future global allergic disease burden looks even more challenging. Changes in allergen exposure will be linked to changes in allergen sensitisation and the prevalence, severity, and distribution of allergic diseases. The spectrum of allergic diseases is broad, including not only asthma and allergic rhinitis (and rhinoconjunctivitis) but also atopic dermatitis, insect sting allergy, food allergy, and others. While the impacts of climate change on allergens are still being researched, the evidence to date suggests many adverse impacts from the perspective of those with allergic disease.

These affects, detailed in the preceding chapters, and summarised here, present an enormous environmental and public health challenge. The final two sections of this chapter (Sections 10.4 and 10.5) explore responses to this challenge. Ahead of these, the following section highlights important knowledge gaps and research needs.

10.3 Knowledge Gaps and Research Needs

Previous chapters of this book (Chapters 2–9) identified a range of knowledge gaps and research needs regarding the aspect of climate change, allergens, and allergic diseases they had focussed on. These are brought together here. Before this though, it will be useful to briefly recall each of these chapters and what their focus was.

Chapter 1 provided a general introduction to the topic and a description of climate change itself – the changes in the composition of the Earth's atmosphere, its temperature, precipitation, and so on. Chapter 2 focussed on the impacts of climate change on aeroallergen (pollen and fungal spore) production and atmospheric concentration. It considered the research that has investigated long-term aerobiological records as well as a range of experimental studies. Chapter 3 examined changes in the types of allergens in our environment by examining the impacts of climate change on the spatial distributions of allergenic species. It explored the evidence for range shifts of allergen-producing plant species as well as a number of stinging insects including wasps, hornets, ants, and bees. Chapter 4 considered the impacts of climate change on the dispersion and transport of environmental aeroallergens within the atmosphere and deposition of them from the atmosphere.

Chapter 5 focussed on changes in allergenicity particularly of pollen but also, to a lesser extent, mould spores, contact allergens, and food allergens. Chapter 6 then examined changes in allergen seasonality, including changes in season start dates, end dates, and durations. Again, this chapter focussed primarily on pollen, on which most of the relevant research has focussed, but also, to a lesser extent, on fungal spores.

The impacts of climate change on indoor allergens were explicitly examined in Chapter 7. The chapter considered house dust mite, cockroach, mouse, and fungi allergens. Chapter 8 tackled the topic of interactions among climate change, air pollutants, and aeroallergens. Following an overview of the impacts of climate change on air pollution, including ozone and particulate matter, the chapter examined the interactions between air pollutants and pollen both within the human body and in the atmosphere.

Finally, Chapter 9 focussed explicitly on the impacts of climate change on allergic diseases themselves. A spectrum of allergic diseases was considered, with asthma and allergic rhinitis occupying much of the coverage, but other important allergic diseases including allergic conjunctivitis, atopic dermatitis, insect sting allergy, and food allergy also being discussed.

What emerges from these chapters is not only an in-depth assessment of what we know about the impacts of climate change on allergens and allergic diseases but also, whether explicitly or implicitly, the fact that there are many gaps in this knowledge and hence an urgent need for further research.

There is a need for more controlled experiments so that the specific role of individual factors (changes in CO_2, temperature, precipitation, humidity, etc.) in climate change impacts on allergens and allergen-producing organisms can be further clarified. More experiments are also required that examine combinations of factors so that the overall influence can be determined. Many more plant species need to be studied experimentally so that our knowledge of the impacts of climate change on plant allergens extends beyond the relatively few plants that have been studied to date. This includes the need to explore the impacts of climate change on plant food allergens. Development of aerobiological networks for climate change research is also a high priority, and because such networks also function to inform management of allergic diseases such as asthma and allergic rhinitis, this development need is detailed in the next section (Section 10.4). More research on the impacts of climate change on fungal spores is required, using experiments and analysis of long-term records where possible. The very recent experimental research by Damialis *et al.* (2015) on several fungal species serves as an excellent example of this.

Much of the research to date has focussed on the Northern Hemisphere, so more research in the future needs to be focussed on the Southern Hemisphere. Further, with much of the research having focussed on North America and Europe, it is important that more research in the future consider climate change impacts on allergens and allergic diseases especially in Africa, South America, China, and India.

While there is much research on species range responses to climate change, as noted in Chapter 3, few studies have explicitly addressed the implications of species range shifts in terms of human allergic diseases. Further assessment is required to fully consider the many factors contributing to species ranges and changes in them.

With regard to aeroallergen dispersion, transport, and deposition, problems currently facing the modelling in this area need to be solved, and research is required in regions other than Europe. Along with a clear need to continue and expand research of allergen seasonality responses to climate change is the requirement to further understand the role vernalisation may play in future tree pollen season changes.

The indoor environment poses many unanswered questions regarding the impacts of climate change on allergens such as house dust mite, cockroach, mouse, and mould. This environment presents a more challenging research prospect for many reasons, including a range of human social, behavioural, and technological factors that come into play. However, given the high percentage of time we spend indoors, and the often very high prevalence of house dust mite allergen sensitisation, research focussing on the impacts of climate change on this organism alone would have great benefit.

Perhaps the greatest need with respect to air pollution and allergen interactions in the context of climate change is for more research on the influence of air pollutants on allergen production in plants and other allergen-producing organisms and interactions between air pollutants and allergens in the atmosphere.

Finally, original research on the impacts of climate change on allergic diseases themselves is rare, so every aspect of this is in urgent need of attention. To highlight just one aspect, our understanding of the impact of outdoor fungal exposures on asthma exacerbations is poor and requires greater study. Associated with this is the need for a much better understanding of threshold levels for sensitisation and symptom elicitation for not only mould but also pollen and other allergens. More generally, relevant environmental and health databases exist that, interrogated explicitly to address the question of climate change–allergic diseases impacts, will likely yield insightful answers.

10.4 Mitigation and Adaptation

Since at least 1990, the global climate change research community has recognised the importance of climate change and the need to develop and put in place a range of response strategies. Such response strategies are categorised as being in one or the other of two fundamental approaches – mitigation and adaptation (see, e.g., IPCC, 1990, 1996). With respect to climate change, mitigation is a 'human intervention to reduce the sources or enhance the sinks of greenhouse gases (GHGs)' (Allwood *et al.*, 2014). Mitigation therefore aims to reduce the extent or magnitude of climate change. In light of interactions among climate change, air pollutants, and aeroallergens, interventions to reduce GHGs in cities via reduced motor vehicle traffic will have co-benefits of also reducing air pollutants (see Chapter 8).

In the context of climate change, adaptation also has a very specific meaning, being 'the process of adjustment to actual or expected climate and its effects' (Agard *et al.*, 2014). Further, in human systems, with human health being one such system, 'adaptation seeks to moderate or avoid harm or exploit beneficial opportunities' (Agard *et al.*, 2014). Like the research on, and our understanding of, the impacts of climate change on allergens and allergic diseases, formulation of realistic adaptation strategies targeted at these impacts has accelerated in recent years (Beggs, 2010, 2015).

Among strategies, perhaps the most important are significant improvements in our monitoring of environmental allergens. One aspect of this is simply the need for more aeroallergen-monitoring sites. English *et al.* (2009), who identified pollen as one of only six environmental health indicators of climate change for the United States, considered that the spatial coverage of the pollen-monitoring stations in the United States is sparse and that it would be preferable to increase their number. A similar case has been made recently for Australia (Beggs *et al.*,

2015), and there are many parts of the world where pollen monitoring is even more spare or entirely absent. While enhanced monitoring in urban areas is important, there is the need for many more monitoring sites in rural and natural areas (Chapter 2; Fernández-Llamazares *et al.*, 2014). Another aspect is the need to complement traditional airborne pollen grain and mould spore monitoring with direct monitoring of airborne allergens. Much has been learned through recent programs in this area in Europe (Cecchi, 2013). Associated with this is the need for continued and better resourcing of pollen-monitoring networks (Beggs *et al.*, 2015; Cecchi, 2013).

Takaro *et al.* (2013) take a more holistic view of monitoring aeroallergens (and allergic respiratory diseases) in the context of climate change, stating that more finely resolved networks of daily pollen-monitoring sites should be 'linked at comparable temporal and spatial scales to reporting networks for near real-time health tracking of allergy and asthma health effects, linked with carbon dioxide emission source data, and other health-relevant air pollutant monitoring' (Takaro *et al.*, 2013).

While national and regional allergen-monitoring perspectives will necessarily continue, our focus must increasingly become more global and international. Aspects of such a focus would include international standardisation of the methods and techniques used and technology transfer to enable poorly serviced population centres, nations, and regions to rapidly establish facilities and services at international standards, especially in developing countries.

Other important adaptation strategies in the context of climate change, allergens, and allergic diseases are improved allergenic plant management and policies, particularly for urban locations (Chapter 8; Beggs, 2010, 2015; Cheng and Berry, 2013; Katz and Carey, 2014). Katz and Carey (2014), for example, in their recent study on ragweed, concluded that 'management of allergenic pollen producing plants must be considered at multiple spatial scales' and 'will become increasingly important over the coming decades, as temperatures and carbon dioxide concentrations continue to increase'.

There are many other adaptation strategies beyond the few mentioned above. Chapter 9, for example, concludes that 'the currently available medical workforce is not meeting the needs of allergic patients in any region, so with the likely changes and increases in some allergic disorders, it is imperative that measures to train and upskill medical workforces in each region be undertaken as a healthcare priority'.

While development and implementation of adaptation strategies targeted at moderating or avoiding the adverse impacts of climate change on allergens and allergic diseases must continue, it is vital that we enhance mitigation action. This is because 'without additional mitigation efforts beyond those in place today, and even with adaptation, warming by the end of the 21st century will lead to high to very high risk of severe, widespread and irreversible impacts globally' (IPCC, 2015).

10.5 A Call to Action

As the previous section has outlined, there are responses, and possible solutions, to the impacts of climate change. As serious and significant as the impacts of climate change on human health are, indeed because such impacts are so significant and serious, *The Lancet* has most recently stated that 'tackling climate change could be the greatest global health opportunity of the 21st century' (Watts *et al.*, 2015).

Our ability to comprehend and project the consequences of anthropogenic climate change on allergens and human health is still in its infancy; there is much more that we need to know. While the gaps in our knowledge make us vulnerable, they also provide the scientific research community with a smorgasbord of important, interesting, relevant, and challenging research opportunities. Such opportunities are there for such researchers at every level – from those embarking on research for the first time to those well established in their careers perhaps looking for a new focus or a new challenge.

The topic requires the expertise in many disciplines: climate sciences, biological sciences (botany, ecology, etc.), environmental sciences, medical and health sciences, allergy and immunology, just to name a few. While these disciplines alone have a lot to offer future inquiry in this area, so too do interdisciplinary researchers: environmental health scientists, aerobiologists, biometeorologists, ecoclimatologists, and so on. Indeed, perhaps the greatest and the most insightful gains will be made through interdisciplinary and even transdisciplinary collaborations (Davies *et al.*, 2015; Lynch *et al.*, 2015).

Research on the impacts of climate change on allergens and allergic diseases continues. That focussing on allergic diseases directly is particularly important given that this aspect of the topic has received less attention than the impacts of climate change on allergens per se. In this regard, the studies by Fuertes *et al.* (2014) and Salo *et al.* (2014) – the former analysing data from the International Study of Asthma and Allergies in Childhood Phase Three, and the latter from the US National Health and Nutrition Examination Survey – serve as excellent examples (Beggs, 2015).

Anthropogenic climate change involves the whole climate system, our atmosphere, hydrosphere, cryosphere, land surface, and biosphere. Allergen-producing organisms are a part of the biosphere, and so are we. We are intimately connected to our environment and the ecosystems of the Earth. Climate change has shown us – perhaps more than ever before – how powerful human actions can be. It is this power that is now required to protect our future health and that of the species and ecosystems with which we share this planet.

References

Agard, J., Schipper, E. L. F., Birkmann, J., *et al.* (2014). Annex II: Glossary. In: Field, C. B., Barros, V. R., Dokken, D. J., *et al.*, eds. *Climate Change 2014: Impacts, Adaptation, and Vulnerability. Part A: Global and Sectoral Aspects. Contribution of Working Group II to the Fifth Assessment Report of the Intergovernmental Panel on Climate Change.* Cambridge, UK and New York, USA: Cambridge University Press, pp. 1757–1776.

Allwood, J. M., Bosetti, V., Dubash, N. K., Gómez-Echeverri, L., von Stechow, C. (2014). Annex I: Glossary, acronyms and chemical symbols. In: Edenhofer, O., Pichs-Madruga, R., Sokona, Y., *et al.*, eds. *Climate Change 2014: Mitigation of Climate Change. Contribution of Working Group III to the Fifth Assessment Report of the Intergovernmental Panel on Climate Change.* Cambridge, UK and New York, USA: Cambridge University Press, pp. 1249–1279.

Beggs, P. J. (2010). Adaptation to impacts of climate change on aeroallergens and allergic respiratory diseases. *International Journal of Environmental Research and Public Health*, 7(8), 3006–3021.

Beggs, P. J. (2015). Environmental allergens: from asthma to hay fever and beyond. *Current Climate Change Reports*, 1(3), 176–184.

Beggs, P. J., Katelaris, C. H., Medek, D., *et al.* (2015). Differences in grass pollen allergen exposure across Australia. *Australian and New Zealand Journal of Public Health*, 39(1), 51–55.

Cecchi, L. (2013). From pollen count to pollen potency: the molecular era of aerobiology. *European Respiratory Journal*, 42(4), 898–900.

Cheng, J. J., Berry, P. (2013). Health co-benefits and risks of public health adaptation strategies to climate change: a review of current literature. *International Journal of Public Health*, 58(2), 305–311.

Damialis, A., Mohammad, A. B., Halley, J. M., Gange, A. C. (2015). Fungi in a changing world: growth rates will be elevated, but spore production may decrease in future climates. *International Journal of Biometeorology*, 59(9), 1157–1167.

Davies, J. M., Beggs, P. J., Medek, D. E., *et al.* (2015). Trans-disciplinary research in synthesis of grass pollen aerobiology and its importance for respiratory health in Australasia. *Science of the Total Environment*, 534, 85–96.

English, P. B., Sinclair, A. H., Ross, Z., *et al.* (2009). Environmental health indicators of climate change for the United States: findings from the State Environmental Health Indicator Collaborative. *Environmental Health Perspectives*, 117(11), 1673–1681.

Fernández-Llamazares, Á., Belmonte, J., Boada, M., Fraixedas, S. (2014). Airborne pollen records and their potential applications to the conservation of biodiversity. *Aerobiologia*, 30(2), 111–122.

Fuertes, E., Butland, B. K., Anderson, H. R., *et al.* (2014). Childhood intermittent and persistent rhinitis prevalence and climate and vegetation: a global ecologic analysis. *Annals of Allergy, Asthma & Immunology*, 113(4), 386–392.

IPCC (1990). *Climate Change: The IPCC Response Strategies. Working Group III of the Intergovernmental Panel on Climate Change* [Bernthal, F., Dowdeswell, E., Luo, J., *et al.*, eds.]. World Meteorological Organization and United Nations Environment Program.

IPCC (1996). *Climate Change 1995: Impacts, Adaptations and Mitigation of Climate Change: Scientific-Technical Analyses. Contribution of Working Group II to the Second Assessment Report of the Intergovernmental Panel on Climate Change* [Watson, R. T., Zinyowera, M. C., Moss, R. H., Dokken, D. J., eds.]. Cambridge, UK and New York, USA: Cambridge University Press.

IPCC (2015). Summary for policymakers. In: The Core Writing Team, Pachauri, R. K., Meyer, L. A., eds. *Climate Change 2014: Synthesis Report. Contribution of Working Groups I, II and III to the Fifth Assessment Report of the Intergovernmental Panel on Climate Change*. Geneva, Switzerland: IPCC, pp. 1–31.

Katz, D. S. W., Carey, T. S. (2014). Heterogeneity in ragweed pollen exposure is determined by plant composition at small spatial scales. *Science of the Total Environment*, 485–486, 435–440.

Longstreth, J. (1991). Anticipated public health consequences of global climate change. *Environmental Health Perspectives*, 96, 139–144.

Lynch, A. J. J., Thackway, R., Specht, A., *et al.* (2015). Transdisciplinary synthesis for ecosystem science, policy and management: the Australian experience. *Science of the Total Environment*, 534, 173–184.

Pawankar, R., Canonica, G. W., Holgate, S. T., Lockey, R. F., eds. (2011). *World Allergy Organization (WAO) White Book on Allergy*. Milwaukee, USA: World Allergy Organization.

Platts-Mills, T. A. E. (2015). The allergy epidemics: 1870–2010. *The Journal of Allergy and Clinical Immunology*, 136(1), 3–13.

Salo, P. M., Arbes Jr, S. J., Jaramillo, R., *et al.* (2014). Prevalence of allergic sensitization in the United States: results from the National Health and Nutrition Examination Survey (NHANES) 2005–2006. *The Journal of Allergy and Clinical Immunology*, 134(2), 350–359.

Takaro, T. K., Knowlton, K., Balmes, J. R. (2013). Climate change and respiratory health: current evidence and knowledge gaps. *Expert Review of Respiratory Medicine*, 7(4), 349–361.

Watts, N., Adger, W. N., Agnolucci, P., *et al.* (2015). Health and climate change: policy responses to protect public health. *The Lancet*, 386(10006), 1861–1914.

Index